G000242066

Yearbook of Astronomy 2021

Front Cover: Launched in 2011 and arriving at Jupiter in 2016, NASA's JUNO spacecraft is currently in a polar orbit that takes it out as far as 8 million kilometres before passing as close as 4,200 kilometres to Jupiter's cloud tops. It is visualised here passing over Jupiter's south pole. More information on the JUNO mission can be found in the articles *Solar System Exploration in 2019* and *Solar System Exploration in 2020* in the corresponding editions of the *Yearbook of Astronomy*. (NASA / JPL-Caltech)

YEARBOOK OF
ASTRONOMY
2021

EDITED BY

Brian Jones

WHITE OWL

AN IMPRINT OF PEN & SWORD BOOKS LTD.
YORKSHIRE - PHILADELPHIA

First published in Great Britain in 2020 by
WHITE OWL
An imprint of
Pen & Sword Books Ltd
Yorkshire – Philadelphia

Copyright © Brian Jones, 2020

ISBN 978 1 52677 187 2

The right of Brian Jones to be identified as Author of this work has been asserted by him in accordance with the Copyright, Designs and Patents Act 1988.

A CIP catalogue record for this book is available from the British Library.

All rights reserved. No part of this book may be reproduced or transmitted in any form or by any means, electronic or mechanical including photocopying, recording or by any information storage and retrieval system, without permission from the Publisher in writing.

Typeset in Dante By Mac Style

Printed and bound by Printworks Global Ltd, London/Hong Kong

Pen & Sword Books Ltd incorporates the Imprints of Pen & Sword Books Archaeology, Atlas, Aviation, Battleground, Discovery, Family History, History, Maritime, Military, Naval, Politics, Railways, Select, Transport, True Crime, Fiction, Frontline Books, Leo Cooper, Praetorian Press, Seaforth Publishing, Wharncliffe and White Owl.

For a complete list of Pen & Sword titles please contact

PEN & SWORD BOOKS LIMITED
47 Church Street, Barnsley, South Yorkshire, S70 2AS, England
E-mail: enquiries@pen-and-sword.co.uk
Website: www.pen-and-sword.co.uk

or

PEN AND SWORD BOOKS
1950 Lawrence Rd, Havertown, PA 19083, USA
E-mail: Uspen-and-sword@casematepublishers.com

Contents

Editor's Foreword

The *Yearbook of Astronomy 2021* is the latest edition of what has long been an indispensable publication, the annual appearance of which has been eagerly anticipated by astronomers, both amateur and professional, for nearly sixty years. Approaching its Diamond Jubilee edition in 2022, the *Yearbook of Astronomy* is, as ever, aimed at both the armchair astronomer and the active backyard observer. Within its pages you will find a rich blend of information, star charts and guides to the night sky coupled with an interesting mixture of articles which collectively embrace a wide range of topics, ranging from the history of astronomy to the latest results of astronomical research; space exploration to observational astronomy; and our own celestial neighbourhood out to the farthest reaches of space.

The *Monthly Star Charts* have been compiled by David Harper and show the night sky as seen throughout the year. Two sets of twelve charts have been provided, one set for observers in the Northern Hemisphere and one for those in the Southern Hemisphere. Between them, each pair of charts depicts the entire sky as two semi-circular half-sky views, one looking north and the other looking south.

Lists of *Phases of the Moon in 2021* and *Eclipses in 2021* are also provided, together with general summaries of the observing conditions for each of the planets in *The Planets in 2021*, and a calendar of significant Solar System events occurring throughout the year in *Some Events in 2021*.

The ongoing process of improving and updating what the *Yearbook of Astronomy* offers to its readers is continued in the 2021 edition with the introduction of apparition charts for Mars, Jupiter and Saturn. All the planetary apparition charts have been compiled by David Harper and the inclusion of the three listed here ensures that charts are now supplied for all the major planetary members of our Solar System. Further details of the planetary apparition charts are given in *Using the Yearbook of Astronomy as an Observing Guide*.

As with *The Planets in 2021* and *Some Events in 2021*, the *Monthly Sky Notes* have been compiled by Lynne Marie Stockman and give details of the positions and visibility of the planets for each month throughout 2021. Each section of the *Monthly Sky Notes* is accompanied by a short article, the range of which includes items on a variety of astronomy-related topics such as *Astronomical Illustrations of the Nuremberg Chronicle* by Richard Sanderson, *2I/Borosov: Interstellar Comet* by Neil Norman and *Burying the Sun* by Carolyn Kennett together with a trio of biographical articles relating to astrophysicist Alfred Fowler, double star observer Thomas Henry Espinell Compton Espin, and one of the best known astronomers of all time, Tycho Brahe.

The Monthly Sky Notes and Articles section of the book concludes with a trio of articles penned by Neil Norman, these being *Comets in 2021*, *Minor Planets in 2021* and *Meteor Showers in 2021*, all three titles being fairly self-explanatory describing as they do the occurrence and visibility of examples of these three classes of object during and throughout the year.

In his article *Astronomy in 2020* Rod Hine covers a range of topics, including taking a look at potential problems revealed by results gleaned from the Planck Observatory, which operated from 2009 to 2013 and which provided extensive data on Cosmic Microwave Background radiation. The problems that seem to have arisen for cosmologists relating to the amount of dark matter in the Universe are examined, and how this may (or may not) be resolved once the Simons Observatory in Chile begins operations in the early-2020s.

This is followed by *Solar System Exploration in 2020* in which Peter Rea covers a wide range of ongoing space missions. These include the collection and return to Earth of samples from the surfaces of two asteroids, one being 162173 Ryugu by the Japanese Hyabusa 2, and the other 101955 Bennu, which has been the target of the American OSIRIS-Rex. The ambitious goals of both these missions highlight the incredible advances that our exploration of other planetary members of the Solar System has made in recent years.

In 2021 we celebrate the 450th anniversary of the birth of Johannes Kepler, best known for his three Laws of Planetary Motion published between 1609 and 1618. Neil Haggath discusses some of Kepler's work in his article *Anniversaries in 2021* as well as drawing to our attention the centenaries of the deaths of several other notable astronomers. These include the 350th anniversary of the death of Italian Jesuit astronomer Father Giovanni Battista Riccioli, who is credited for introducing the system of lunar nomenclature still in use today; the 150th anniversary of the death of Sir John Herschel, son of the great observer

Sir William Herschel; and the centenary of the death of American astronomer double star observer Sherburne Wesley Burnham, who is best known for the discovery of 1,340 double stars and for the publication of the Burnham Double Star Catalogue, which listed 13,665 binaries.

The article *Mission to Mars: Countdown to Building a Brave New World* by Martin Braddock is the first in a series of articles scheduled to appear in the Yearbook of Astronomy throughout the 2020s and which will keep the reader fully up to date with ongoing preparations that will have been conducted so far. New research and preparations are currently underway and in the planning phase, all of which are geared towards sending a manned mission to Mars at or around the end of the decade. We are at the start of what promises to be an exciting journey.

The article *Male Family Mentors for Women in Astronomy: En'hedu'anna to Eimmart* is the first of three articles by Mary McIntyre devoted to examining the work and achievements of famous and inspirational women astronomers who had the support of a male partner or family member. Notable among the numerous female astronomers covered in the article is Sophia Brahe, the younger sister of Danish nobleman and astronomer Tycho Brahe. Sophia had developed a love of the stars during early childhood and carried her enthusiasm through to become a capable and skilful assistant to her elder brother. Among the other female astronomers discussed is Elisabetha Hevelius, the second wife of famous astronomer Johannes Hevelius. Elisabetha took on the role of assistant to Johannes, performing her own observations and undertaking important mathematical calculations. The second article of the series, due for publication in the forthcoming *Yearbook of Astronomy 2023*, will focus on the brother-sister astronomy partnership of William and Caroline Herschel.

In *Henrietta Swan Leavitt and Her Work* by regular contributor David M. Harland we learn about the achievements of Henrietta Swan Leavitt. Originally hired by Harvard College Observatory director Edward Charles Pickering as a female assistant, or 'computer', to assist in the classification of stellar spectra, she is best known for her discovery of the Cepheid variable period-luminosity relationship. This ground-breaking achievement was destined to help other astronomers make further notable discoveries, eventually revealing the vast size of the universe.

As the title suggests, *Solar Observing* by Peter Meadows takes us through the various methods and techniques used, and the equipment needed, in order to carry out observations of our nearest star, the Sun. As we learn from the article, the Sun is a fascinating object for the amateur astronomer and presents the observer with something different to see each day. In *The Meteorite Age* by Carolyn Kennett, we are given a fascinating account of the discovery of worked pieces of meteoritic iron dating back many thousands of years. As we learn from the article, the presence of these iron objects, which have been found to predate the Iron Age, suggests to us that meteorite objects may represent the earliest encounters that humans had with metal.

Although the constellations we see on our star charts now seem fixed and ancient and the sky appears orderly, this has not always been the case. In his article *'A Dignity That Insures Their Perpetuation'* John Barentine explains how constellations were once recognized, relates how the modern night sky came to be, and describes the figures that were discarded along the way.

Although it was once believed that many lunar craters were the result of volcanic activity, we now recognise them as having being formed by the impacts of asteroids and meteoroids. However, many images of the lunar surface obtained by orbiting spacecraft have revealed features resembling large lava flows. In his article *Lunar Volcanism* Lionel Wilson tells us how volcanic activity has played an important part in the formation of the lunar surface we see today.

In his *Pages From the Past: Collecting Vintage Astronomy Books* Richard Sanderson tells us of his passion for collecting vintage astronomy books. Bringing together personal anecdotes gleaned from decades of book sleuthing with information and suggestions for those new to the hobby, this article will certainly encourage many readers to take up and embrace this rewarding intellectual activity.

It is definitely a case of 'from little green men to microbes' in *The Chances of Anything Coming From Mars* by Jan Hardy, in which she looks at how the observations of 'canali' on Mars by eighteenth-century Italian astronomer Giovanni Schiaparelli went on to influence the depiction of the red planet in science fiction as home to alien civilisations, with often hostile ambitions!

Following on from the article relating to Aboriginal astronomy in the 2020 edition of the Yearbook of Astronomy, *Māori Astronomy in Aotearoa-New Zealand* by Pauline Harris, Hēmi Whaanga and Rangi Matamua tells of the ongoing and significant efforts of the Māori movement to reclaim their astronomical knowledge. We see how this revitalisation of astronomical understanding has received a great deal of support from both Māori and non-Māori communities. In addition, we are given an insight into how the heavens were perceived by ancestors of the Māori, as well as learning how the Māori perception of the Sun, Moon and stars played an integral role in their agriculture, architecture, fishing and exploration.

The final section of the book starts off with *Some Interesting Variable Stars* by Tracie Heywood which contains useful information on variables as well as predictions for timings of minimum brightness of the famous eclipsing binary Algol for 2021. *Some Interesting Double Stars* and *Some Interesting Nebulae, Star Clusters and Galaxies* present a selection of objects for you to seek out in the night sky. The lists included here are by no means definitive and may well omit your favourite celestial targets. If this is the case, please let us know and we will endeavour to include these in future editions of the Yearbook.

The book rounds off with a selection of *Astronomical Organizations*, which lists organizations and associations across the world through which you can further pursue your interest and participation in astronomy (if there are any that we have omitted please let us know) and *Our Contributors*, which contains brief background details of the numerous writers who have contributed to this edition of the Yearbook. The book rounds off with the *Glossary*, which benefits from the addition of new entries of brief but informative explanations for words and terminology used in this and previous editions of the Yearbook.

Over time new topics and themes will be introduced into the Yearbook to allow it to keep pace with the increasing range of skills, techniques and observing methods now open to amateur astronomers, this in addition to articles relating to our rapidly-expanding knowledge of the Universe in which we live. There will be an interesting mix, some articles written at a level which will appeal to the casual reader and some of what may be loosely described as at a more academic level. The intention is to fully maintain and continually increase the usefulness and relevance of the Yearbook of Astronomy to the interests of the readership who are, without doubt, the most important aspect of the Yearbook and the reason it exists in the first place. With this in mind, suggestions from readers for further improvements and additions to the Yearbook are always welcomed. After all, the book is written for you …

As ever, grateful thanks are extended to those individuals who have contributed a great deal of time and effort to the *Yearbook of Astronomy 2021*, including David Harper, who has provided updated versions of his excellent Monthly Star Charts. These were generated specifically for what has been described as the new generation of the Yearbook of Astronomy, and the charts add greatly to the overall value of the book to star gazers. Equally important are the efforts of Lynne Marie Stockman who has put together the Monthly Sky Notes. Their combined efforts have produced what can justifiably be described as the backbone of the Yearbook of Astronomy.

In addition to the above I would like to thank Roger Pickard for his articles *Some Interesting Variable Stars* written for the 2018, 2019 and 2020 editions of the Yearbook of Astronomy. His efforts in the past have certainly helped this section of the Yearbook to become an extremely useful observing guide for the many variable star observers who read the book, a role in which we are confident that new contributor Tracie Heywood will succeed!

Also worthy of mention are Mat Blurton, who has done an excellent job typesetting the Yearbook, and Jonathan Wright, Emily Robinson, Lori Jones, Janet Brookes and Paul Wilkinson of Pen & Sword Books Ltd without whose combined help, belief and confidence in the Yearbook of Astronomy, this much-loved and iconic publication may well have disappeared for ever.

Brian Jones – Editor
Bradford, West Riding of Yorkshire

February 2020

As many of you will be aware, the future of the Yearbook of Astronomy was under threat following the decision to make the 2016 edition the last. However, the series was rescued, both through the publication of a special 2017 edition (which successfully maintained the continuity of the Yearbook) and a successful search for a publisher to take this iconic publication on and to carry it to even greater heights as the Yearbook approaches its Diamond Jubilee in 2022.

The *Yearbook of Astronomy 2017* was a limited edition, although copies are still available to purchase. It should be borne in mind that you would not be obtaining the 2017 edition as a current guide to the night sky, but as the landmark edition of the Yearbook of Astronomy which fulfilled its purpose of keeping the series alive, and which heralded in the new generation of this highly valued and treasured publication. You can order your copy of the 2017 edition at **www.starlight-nights.co.uk/subscriber-2017-yearbook-astronomy**

Preface

The information given in this edition of the Yearbook of Astronomy is in narrative form. The positions of the planets given in the Monthly Sky Notes often refer to the constellations in which they lie at the time. These can be found on the star charts which collectively show the whole sky via two charts depicting the northern and southern circumpolar stars and forty-eight charts depicting the main stars and constellations for each month of the year. The northern and southern circumpolar charts show the stars that are within 45° of the two celestial poles, while the monthly charts depict the stars and constellations that are visible throughout the year from Europe and North America or from Australia and New Zealand. The monthly charts overlap the circumpolar charts. Wherever you are on the Earth, you will be able to locate and identify the stars depicted on the appropriate areas of the chart(s).

There are numerous star atlases available that offer more detailed information, such as Sky & Telescope's POCKET SKY ATLAS and Norton's STAR ATLAS and Reference Handbook to name but a couple. In addition, more precise information relating to planetary positions and so on can be found in a number of publications, a good example of which is The Handbook of the British Astronomical Association, as well as many of the popular astronomy magazines such as the British monthly periodicals Sky at Night and Astronomy Now and the American monthly magazines Astronomy and Sky & Telescope.

About Time

Before the late eighteenth century, the biggest problem affecting mariners sailing the seas was finding their position. Latitude was easily determined by observing the altitude of the pole star above the northern horizon. Longitude, however, was far more difficult to measure. The inability of mariners to determine their longitude often led to them getting lost, and on many occasions shipwrecked. To address this problem King Charles II established the Royal Observatory at Greenwich in 1675 and from here, Astronomers Royal began the process of measuring and cataloguing the stars as they passed due south across the Greenwich meridian.

Now mariners only needed an accurate timepiece (the chronometer invented by Yorkshire-born clockmaker John Harrison) to display GMT (Greenwich Mean Time). Working out the local standard time onboard ship and subtracting this from GMT gave the ship's longitude (west or east) from the Greenwich meridian. Therefore mariners always knew where they were at sea and the longitude problem was solved.

Astronomers use a time scale called Universal Time (UT). This is equivalent to Greenwich Mean Time and is defined by the rotation of the Earth. The Yearbook of Astronomy gives all times in UT rather than in the local time for a particular city or country. Times are expressed using the 24-hour clock, with the day beginning at midnight, denoted by 00:00. Universal Time (UT) is related to local mean time by the formula:

Local Mean Time = UT – west longitude

In practice, small differences in longitude are ignored and the observer will use local clock time which will be the appropriate Standard (or Zone) Time. As the formula indicates, places in west longitude will have a Standard Time slow on UT, while those in east longitude will have a Standard Time fast on UT. As examples we have:

Standard Time in

New Zealand	UT +12 hours
Victoria, NSW	UT +10 hours
Western Australia	UT + 8 hours
South Africa	UT + 2 hours
British Isles	UT
Eastern Standard Time	UT −5 hours
Central Standard Time	UT −6 hours
Pacific Standard Time	UT −8 hours

During the periods when Summer Time (also called Daylight Saving Time) is in use, one hour must be added to Standard Time to obtain the appropriate Summer/Daylight Saving Time. For example, Pacific Daylight Time is UT −7 hours.

Using the Yearbook of Astronomy as an Observing Guide

Notes on the Monthly Star Charts

The star charts on the following pages show the night sky throughout the year. There are two sets of charts, one for use by observers in the Northern Hemisphere and one for those in the Southern Hemisphere. The first set is drawn for latitude 52°N and can be used by observers in Europe, Canada and most of the United States. The second set is drawn for latitude 35°S and show the stars as seen from Australia and New Zealand. Twelve pairs of charts are provided for each of these latitudes.

Each pair of charts shows the entire sky as two semi-circular half-sky views, one looking north and the other looking south. A given pair of charts can be used at different times of year. For example, chart 1 shows the night sky at midnight on 21 December, but also at 2am on 21 January, 4am on 21 February and so forth. The accompanying table will enable you to select the correct chart for a given month and time of night. The caption next to each chart also lists the dates and times of night for which it is valid.

The charts are intended to help you find the more prominent constellations and other objects of interest mentioned in the monthly observing notes. To avoid the charts becoming too crowded, only stars of magnitude 4.5 or brighter are shown. This corresponds to stars that are bright enough to be seen from any dark suburban garden on a night when the Moon is not too close to full phase.

Each constellation is depicted by joining selected stars with lines to form a pattern. There is no official standard for these patterns, so you may occasionally find different patterns used in other popular astronomy books for some of the constellations.

Any map projection from a sphere onto a flat page will by necessity contain some distortions. This is true of star charts as well as maps of the Earth. The distortion on the half-sky charts is greatest near the semi-circular boundary of each chart, where it may appear to stretch constellation patterns out of shape.

The charts also show selected deep-sky objects such as galaxies, nebulae and star clusters. Many of these objects are too faint to be seen with the naked eye, and you will need binoculars or a telescope to observe them. Please refer to the table of deep-sky objects for more information.

Planetary Apparition Diagrams

The diagrams of the apparitions of Mercury and Venus show the position of the respective planet in the sky at the moment of sunrise or sunset throughout the entire apparition. Two sets of positions are plotted on each chart: for latitude 52° North (blue line) and for latitude 35° South (red line). A thin dotted line denotes the portion of the apparition which falls outside the year covered by this edition of the Yearbook. A white dot indicates the position of Venus on the first day of each month, or of Mercury on the first, eleventh and 21st of the month. The day of greatest elongation (GE) is also marked by a white dot. Note that the dots do NOT indicate the magnitude of the planet.

The finder chart for Mars shows its path during the first half of 2021, when it is an evening object as it moves away from opposition in October of last year. Mars traverses more than 100° in ecliptic longitude during this period, moving from Pisces in January to Cancer in June, so the chart is based upon the ecliptic, which runs across the centre of the chart from right to left. The position of Mars is indicated on the 1st of each month from January to July. Stars are shown to magnitude 5.5. Note that the sizes of the Mars dots do NOT indicate its magnitude.

The finder charts for Jupiter, Saturn, Uranus and Neptune show the paths of the planets throughout the year. The position of each planet is indicated at opposition and at stationary points, as well as the start and end of the year and on the 1st of each month (1st of April, July and October only for Uranus and Neptune) where these dates do not fall too close to an event that is already marked. Stars are shown to magnitude 5.5 on the charts for Jupiter and Saturn. On the Uranus chart, stars are shown to magnitude 8; on the Neptune chart, the limiting magnitude is 10. In both cases, this is approximately two magnitudes fainter than the planet itself. Right Ascension and Declination scales are shown for the epoch J2000 to allow comparison with modern star charts. Note that the sizes of the dots denoting the planets do NOT indicate their magnitudes.

Selecting the Correct Charts

The table below shows which of the charts to use for particular dates and times throughout the year and will help you to select the correct pair of half-sky charts for any combination of month and time of night.

The Earth takes 23 hours 56 minutes (and 4 seconds) to rotate once around its axis with respect to the fixed stars. Because this is around four minutes shorter than a full 24 hours, the stars appear to rise and set about 4 minutes earlier on each successive day, or around an hour earlier each fortnight. Therefore, as well as showing the stars at 10pm (22h in 24-hour notation) on 21 January, chart 1 also depicts the sky at 9pm (21h) on 6 February, 8pm (20h) on 21 February and 7pm (19h) on 6 March.

The times listed do not include summer time (daylight saving time), so if summer time is in force you must subtract one hour to obtain standard time (GMT if you are in the United Kingdom) before referring to the chart. For example, to find the correct chart for mid-September in the northern hemisphere at 3am summer time, first of all subtract one hour to obtain 2am (2h) standard time. Then you can consult the table, where you will find that you should use chart 11.

The table does not indicate sunrise, sunset or twilight. In northern temperate latitudes, the sky is still light at 18h and 6h from April to September, and still light at 20h and 4h from May to August. In Australia and New Zealand, the sky is still light at 18h and 6h from October to March, and in twilight (with only bright stars visible) at 20h and 04h from November to January.

Local Time	18h	20h	22h	0h	2h	4h	6h
January	11	12	1	2	3	4	5
February	12	1	2	3	4	5	6
March	1	2	3	4	5	6	7
April	2	3	4	5	6	7	8
May	3	4	5	6	7	8	9
June	4	5	6	7	8	9	10
July	5	6	7	8	9	10	11
August	6	7	8	9	10	11	12
September	7	8	9	10	11	12	1
October	8	9	10	11	12	1	2
November	9	10	11	12	1	2	3
December	10	11	12	1	2	3	4

Legend to the Star Charts

STARS		DEEP-SKY OBJECTS	
Symbol	Magnitude	Symbol	Type of object
•	0 or brighter	⁂	Open star cluster
•	1	○	Globular star cluster
•	2	□	Nebula
•	3	▣	Cluster with nebula
•	4	○	Planetary nebula
·	5	◌	Galaxy
✦	Double star		Magellanic Clouds
⊙	Variable star		

Star Names

There are over 200 stars with proper names, most of which are of Roman, Greek or Arabic origin although only a couple of dozen or so of these names are used regularly. Examples include Arcturus in Boötes, Castor and Pollux in Gemini and Rigel in Orion.

A system whereby Greek letters were assigned to stars was introduced by the German astronomer and celestial cartographer Johann Bayer in his star atlas Uranometria, published in 1603. Bayer's system is applied to the brighter stars within any particular constellation, which are given a letter from the Greek alphabet followed by the genitive case of the constellation in which the star is located. This genitive case is simply the Latin form meaning 'of' the constellation. Examples are the stars Alpha Boötis and Beta Centauri which translate literally as 'Alpha of Boötes' and 'Beta of the Centaur'.

As a general rule, the brightest star in a constellation is labelled Alpha (α), the second brightest Beta (β), and the third brightest Gamma (γ) and so on, although there are some constellations where the system falls down. An example is Gemini where the principal star (Pollux) is designated Beta Geminorum, the second brightest (Castor) being known as Alpha Geminorum.

There are only 24 letters in the Greek alphabet, the consequence of which was that the fainter naked eye stars needed an alternative system of classification. The system in popular use is that devised by the first Astronomer Royal John

Flamsteed in which the stars in each constellation are listed numerically in order from west to east. Although many of the brighter stars within any particular constellation will have both Greek letters and Flamsteed numbers, the latter are generally used only when a star does not have a Greek letter.

The Greek Alphabet

α	Alpha	ι	Iota	ρ	Rho
β	Beta	κ	Kappa	σ	Sigma
γ	Gamma	λ	Lambda	τ	Tau
δ	Delta	μ	Mu	υ	Upsilon
ε	Epsilon	ν	Nu	φ	Phi
ζ	Zeta	ξ	Xi	χ	Chi
η	Eta	ο	Omicron	ψ	Psi
θ	Theta	π	Pi	ω	Omega

The Names of the Constellations

On clear, dark, moonless nights, the sky seems to teem with stars although in reality you can never see more than a couple of thousand or so at any one time when looking with the unaided eye. Each and every one of these stars belongs to a particular constellation, although the constellations that we see in the sky, and which grace the pages of star atlases, are nothing more than chance alignments. The stars that make up the constellations are often situated at vastly differing distances from us and only appear close to each other, and form the patterns that we see, because they lie in more or less the same direction as each other as seen from Earth.

A large number of the constellations are named after mythological characters, and were given their names thousands of years ago. However, those star groups lying close to the south celestial pole were discovered by Europeans only during the last few centuries, many of these by explorers and astronomers who mapped the stars during their journeys to lands under southern skies. This resulted in many of the newer constellations having modern-sounding names, such as Octans (the Octant) and Microscopium (the Microscope), both of which were devised by the French astronomer Nicolas Louis De La Caille during the early 1750s.

Over the centuries, many different suggestions for new constellations have been put forward by astronomers who, for one reason or another, felt the need to add new groupings to star charts and to fill gaps between the traditional constellations. Astronomers drew up their own charts of the sky, incorporating their new groups into them. A number of these new constellations had cumbersome names, notable examples including Officina Typographica (the Printing Shop) introduced by the German astronomer Johann Bode in 1801; Sceptrum Brandenburgicum (the Sceptre of Brandenburg) introduced by the German astronomer Gottfried Kirch in 1688; Taurus Poniatovii (Poniatowski's Bull) introduced by the Polish-Lithuanian astronomer Martin Odlanicky Poczobut in 1777; and Quadrans Muralis (the Mural Quadrant) devised by the French astronomer Joseph-Jerôme de Lalande in1795. Although these have long since been rejected, the latter has been immortalised by the annual Quadrantid meteor shower, the radiant of which lies in an area of sky formerly occupied by Quadrans Muralis.

During the 1920s the International Astronomical Union (IAU) systemised matters by adopting an official list of 88 accepted constellations, each with official spellings and abbreviations. Precise boundaries for each constellation were then drawn up so that every point in the sky belonged to a particular constellation.

The abbreviations devised by the IAU each have three letters which in the majority of cases are the first three letters of the constellation name, such as AND for Andromeda, EQU for Equuleus, HER for Hercules, ORI for Orion and so on. This trend is not strictly adhered to in cases where confusion may arise. This happens with the two constellations Leo (abbreviated LEO) and Leo Minor (abbreviated LMI). Similarly, because Triangulum (TRI) may be mistaken for Triangulum Australe, the latter is abbreviated TRA. Other instances occur with Sagitta (SGE) and Sagittarius (SGR) and with Canis Major (CMA) and Canis Minor (CMI) where the first two letters from the second names of the constellations are used. This is also the case with Corona Australis (CRA) and Corona Borealis (CRB) where the first letter of the second name of each constellation is incorporated. Finally, mention must be made of Crater (CRT) which has been abbreviated in such a way as to avoid confusion with the aforementioned CRA (Corona Australis).

The table shown on the following pages contains the name of each of the 88 constellations together with the translation and abbreviation of the constellation name. The constellations depicted on the monthly star charts are identified with their abbreviations rather than the full constellation names.

The Constellations

Andromeda	Andromeda	AND		Delphinus	The Dolphin	DEL
Antlia	The Air Pump	ANT		Dorado	The Goldfish	DOR
Apus	The Bird of Paradise	APS		Draco	The Dragon	DRA
				Equuleus	The Foal	EQU
Aquarius	The Water Carrier	AQR		Eridanus	The River	ERI
Aquila	The Eagle	AQL		Fornax	The Furnace	FOR
Ara	The Altar	ARA		Gemini	The Twins	GEM
Aries	The Ram	ARI		Grus	The Crane	GRU
Auriga	The Charioteer	AUR		Hercules	Hercules	HER
Boötes	The Herdsman	BOO		Horologium	The Pendulum Clock	HOR
Caelum	The Graving Tool	CAE				
Camelopardalis	The Giraffe	CAM		Hydra	The Water Snake	HYA
Cancer	The Crab	CNC		Hydrus	The Lesser Water Snake	HYI
Canes Venatici	The Hunting Dogs	CVN		Indus	The Indian	IND
Canis Major	The Great Dog	CMA		Lacerta	The Lizard	LAC
Canis Minor	The Little Dog	CMI		Leo	The Lion	LEO
Capricornus	The Goat	CAP		Leo Minor	The Lesser Lion	LMI
Carina	The Keel	CAR		Lepus	The Hare	LEP
Cassiopeia	Cassiopeia	CAS		Libra	The Scales	LIB
Centaurus	The Centaur	CEN		Lupus	The Wolf	LUP
Cepheus	Cepheus	CEP		Lynx	The Lynx	LYN
Cetus	The Whale	CET		Lyra	The Lyre	LYR
Chamaeleon	The Chameleon	CHA		Mensa	The Table Mountain	MEN
Circinus	The Pair of Compasses	CIR		Microscopium	The Microscope	MIC
Columba	The Dove	COL		Monoceros	The Unicorn	MON
Coma Berenices	Berenice's Hair	COM		Musca	The Fly	MUS
Corona Australis	The Southern Crown	CRA		Norma	The Level	NOR
Corona Borealis	The Northern Crown	CRB		Octans	The Octant	OCT
				Ophiuchus	The Serpent Bearer	OPH
Corvus	The Crow	CRV		Orion	Orion	ORI
Crater	The Cup	CRT		Pavo	The Peacock	PAV
Crux	The Cross	CRU		Pegasus	Pegasus	PEG
Cygnus	The Swan	CYG		Perseus	Perseus	PER

Phoenix	The Phoenix	PHE	Sextans	The Sextant	SEX
Pictor	The Painter's Easel	PIC	Taurus	The Bull	TAU
Pisces	The Fish	PSC	Telescopium	The Telescope	TEL
Piscis Austrinus	The Southern Fish	PSA	Triangulum	The Triangle	TRI
Puppis	The Stern	PUP	Triangulum Australe	The Southern Triangle	TRA
Pyxis	The Mariner's Compass	PYX	Tucana	The Toucan	TUC
Reticulum	The Net	RET	Ursa Major	The Great Bear	UMA
Sagitta	The Arrow	SGE	Ursa Minor	The Little Bear	UMI
Sagittarius	The Archer	SGR	Vela	The Sail	VEL
Scorpius	The Scorpion	SCO	Virgo	The Virgin	VIR
Sculptor	The Sculptor	SCL	Volans	The Flying Fish	VOL
Scutum	The Shield	SCT	Vulpecula	The Fox	VUL
Serpens Caput and Cauda	The Serpent	SER			

The Monthly Star Charts

Northern Hemisphere Star Charts

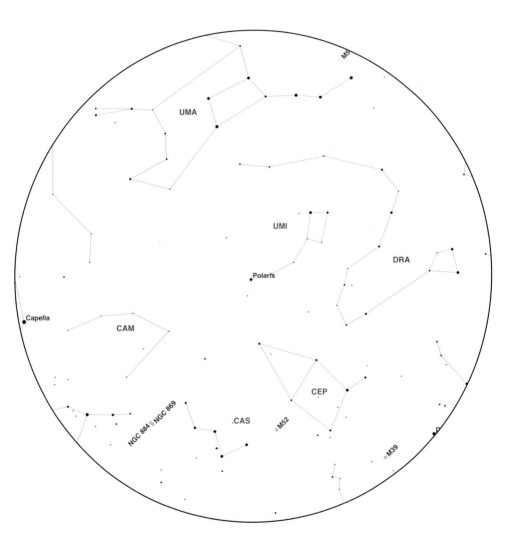

This chart shows stars lying at declinations between +45 and +90 degrees. These constellations are circumpolar for observers in Europe and North America.

1N

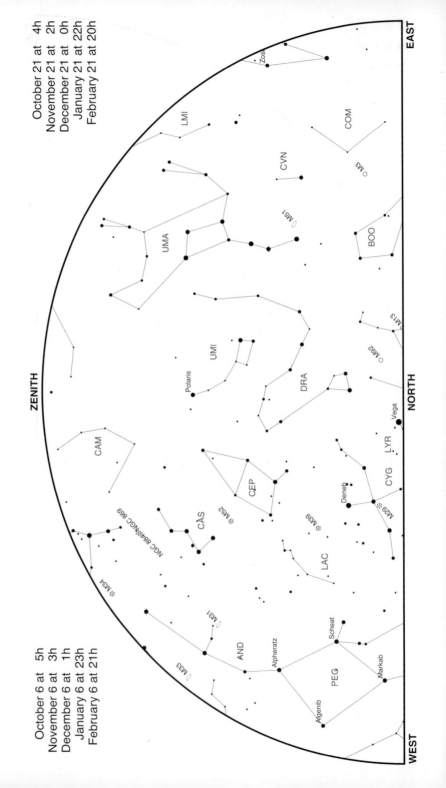

October 21 at 4h
November 21 at 2h
December 21 at 0h
January 21 at 22h
February 21 at 20h

October 6 at 5h
November 6 at 3h
December 6 at 1h
January 6 at 23h
February 6 at 21h

EAST

ZENITH

NORTH

WEST

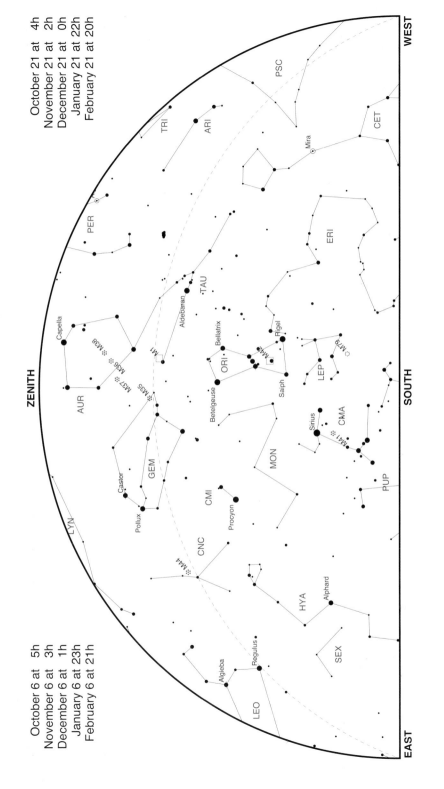

1S

WEST

ZENITH

SOUTH

EAST

October 21 at 4h
November 21 at 2h
December 21 at 0h
January 21 at 22h
February 21 at 20h

October 6 at 5h
November 6 at 3h
December 6 at 1h
January 6 at 23h
February 6 at 21h

PSC
CET
Mira
TRI
ARI
PER
ERI
TAU
Aldebaran
Capella
M38
M36
M37
M35
M41
AUR
Bellatrix
Rigel
ORI
M42
Betelgeuse
Saiph
LEP
M79
CMA
Sirius
M41
GEM
Castor
Pollux
MON
PUP
LYN
CMI
Procyon
CNC
M44
HYA
Alphard
SEX
Regulus
Algieba
LEO

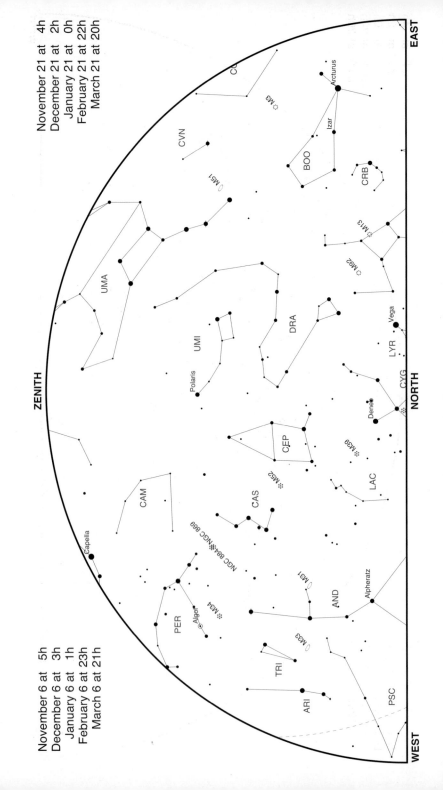

ZENITH

EAST

NORTH

WEST

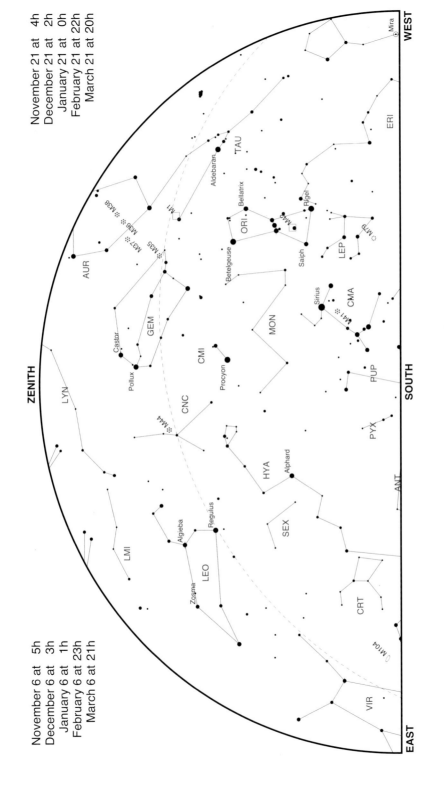

2S

WEST

November 21 at 4h
December 21 at 2h
January 21 at 0h
February 21 at 22h
March 21 at 20h

November 6 at 5h
December 6 at 3h
January 6 at 1h
February 6 at 23h
March 6 at 21h

ZENITH

EAST

SOUTH

3N

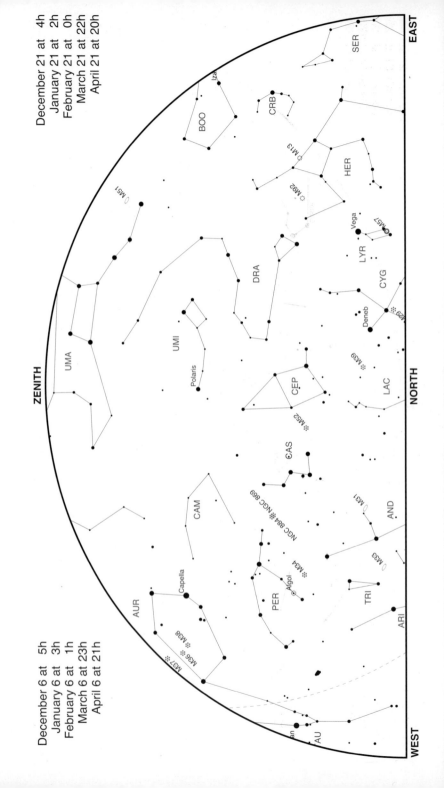

December 21 at 4h
January 21 at 2h
February 21 at 0h
March 21 at 22h
April 21 at 20h

December 6 at 5h
January 6 at 3h
February 6 at 1h
March 6 at 23h
April 6 at 21h

EAST

WEST

NORTH

ZENITH

SER

CRB

BOO

Izar

HER

M13

M92

M57

Vega

LYR

CYG

Deneb

M29

M39

LAC

M31

DRA

M51

UMI

Polaris

UMA

CEP

M52

CAS

CAM

NGC 884 NGC 869

M34

AND

M33

AUR

Capella

M38

M36

M37

PER

Algol

TRI

ARI

AU

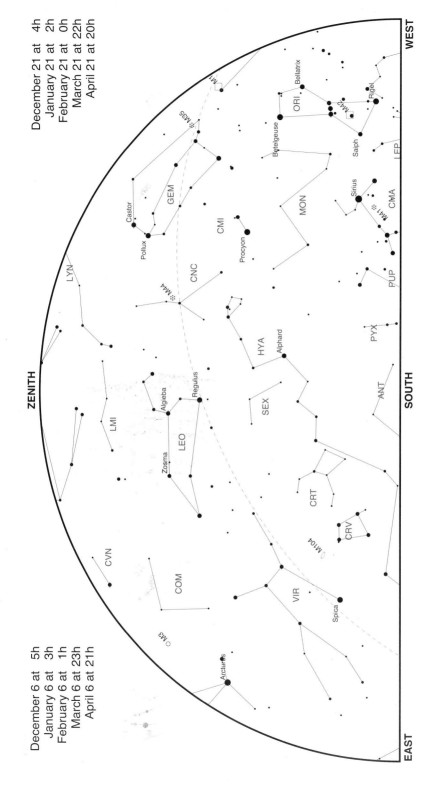

3S

WEST

EAST

SOUTH

ZENITH

Castor
Pollux
GEM
LYN
M35
ORI
Bellatrix
Rigel
M42
Saiph
Betelgeuse
LEP
Sirius
M41
CMA
MON
CMI
Procyon
CNC
M44
PUP
PYX
HYA
Alphard
ANT
SEX
LMI
Algieba
Regulus
LEO
Zosma
CVN
COM
M3
Arcturus
VIR
Spica
M104
CRT
CRV

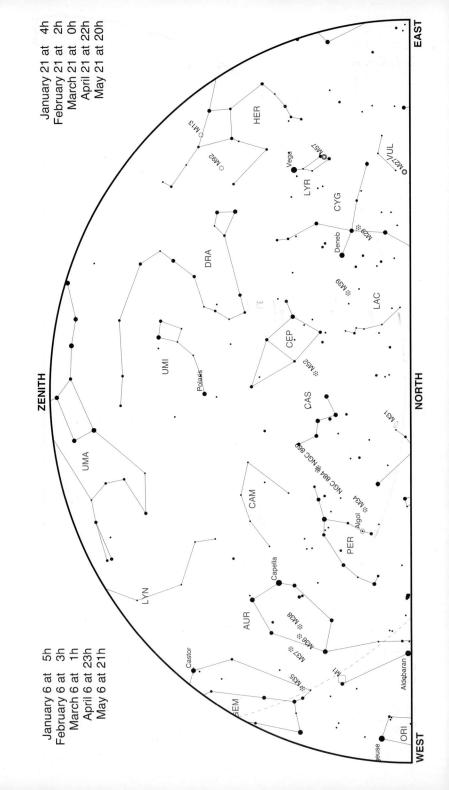

4N

January 21 at 4h
February 21 at 2h
March 21 at 0h
April 21 at 22h
May 21 at 20h

January 6 at 5h
February 6 at 3h
March 6 at 1h
April 6 at 23h
May 6 at 21h

EAST

WEST

ZENITH

NORTH

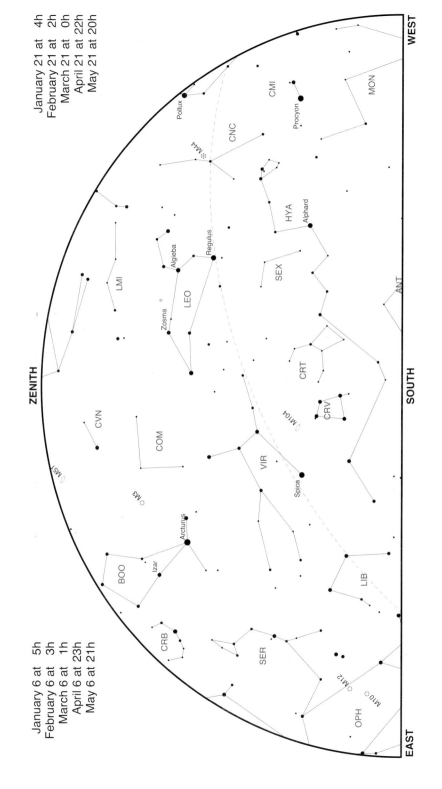

5N

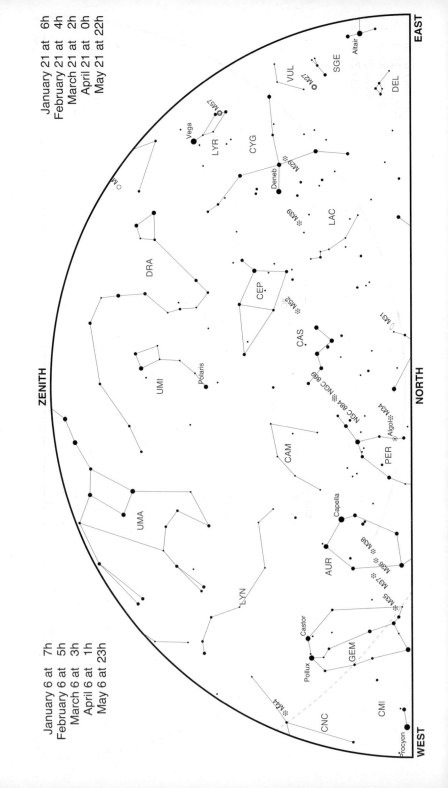

January 6 at 7h
February 6 at 5h
March 6 at 3h
April 6 at 1h
May 6 at 23h

EAST

ZENITH

NORTH

WEST

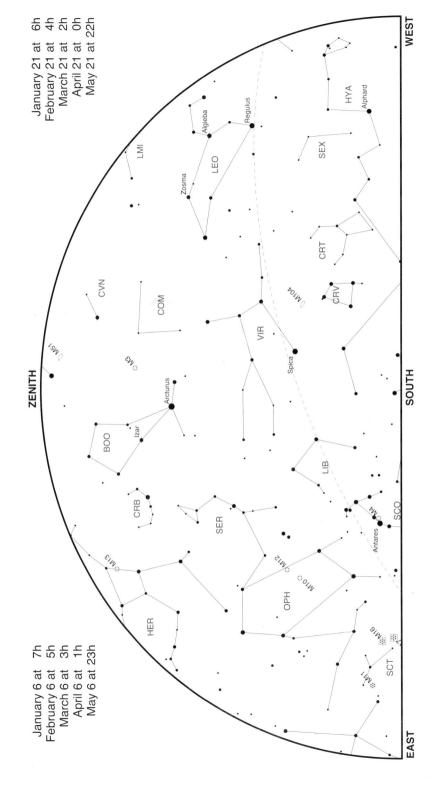

WEST

January 21 at 6h
February 21 at 4h
March 21 at 2h
April 21 at 0h
May 21 at 22h

HYA
Alphard
SEX
Algieba
Regulus
LEO
LMI
Zosma
CRT
M104
CRV
VIR
Spica
CVN
COM
M3
ZENITH
M51
Arcturus
Izar
BOO
LIB
SOUTH
CRB
SER
M4
SCO
Antares
M13
M12
M10
OPH
HER
M16
SCT
M11

January 6 at 7h
February 6 at 5h
March 6 at 3h
April 6 at 1h
May 6 at 23h

EAST

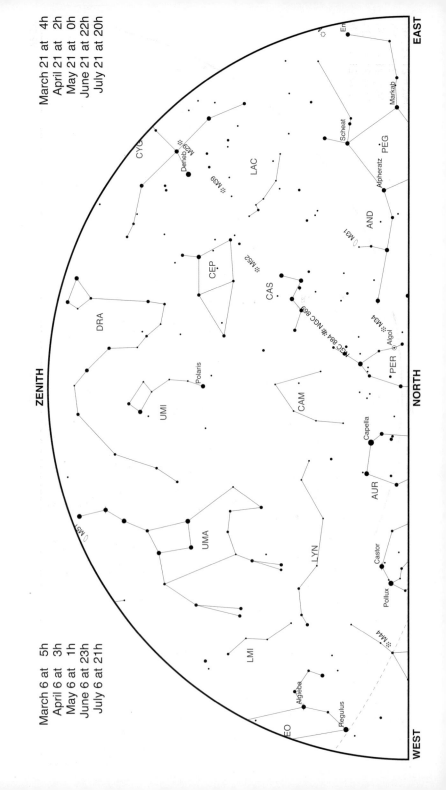

6N

March 6 at 5h
April 6 at 3h
May 6 at 1h
June 6 at 23h
July 6 at 21h

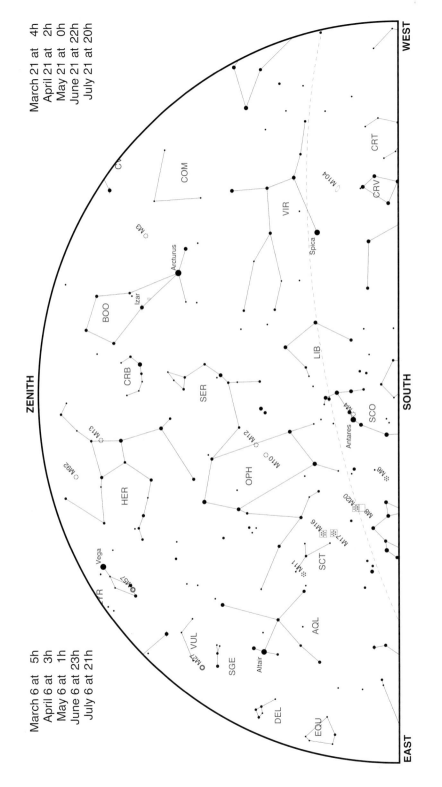

6S

WEST

March 21 at 4h
April 21 at 2h
May 21 at 0h
June 21 at 22h
July 21 at 20h

March 6 at 5h
April 6 at 3h
May 6 at 1h
June 6 at 23h
July 6 at 21h

ZENITH

EAST

SOUTH

CV

COM

VIR

M104

CRT

CRV

Spica

Arcturus

Izar

BOO

M3

LIB

CRB

SER

M13

M92

HER

OPH

M12

M10

M4

Antares

SCO

M6

M8 M20

M17 M16

M11

SCT

M57

Vega

LYR

VUL

M27

SGE

Altair

AQL

DEL

EQU

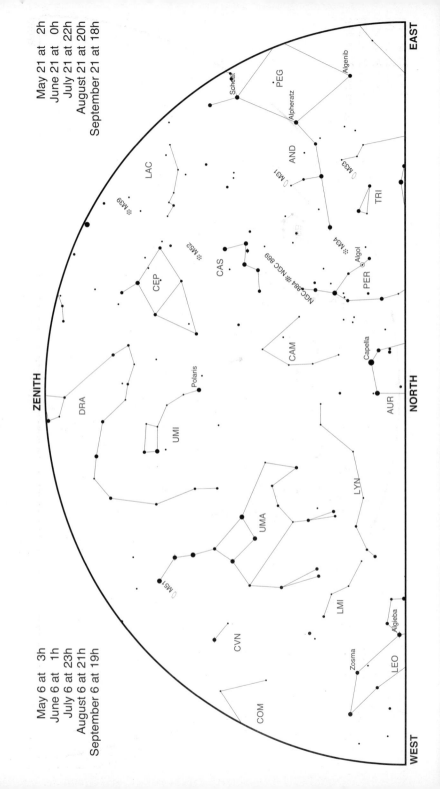

7N

May 6 at 3h
June 6 at 1h
July 6 at 23h
August 6 at 21h
September 6 at 19h

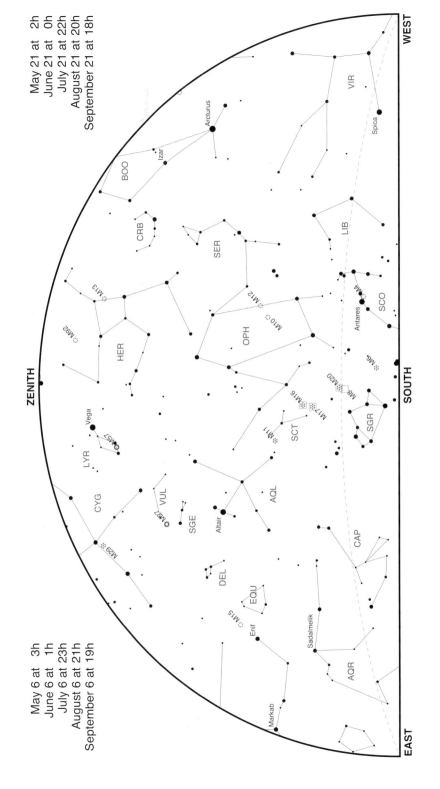

7S

WEST

May 21 at 2h
June 21 at 0h
July 21 at 22h
August 21 at 20h
September 21 at 18h

May 6 at 3h
June 6 at 1h
July 6 at 23h
August 6 at 21h
September 6 at 19h

ZENITH

SOUTH

EAST

Arcturus
Spica
VIR
BOO
Izar
CRB
SER
LIB
M13
M92
M12
M10
HER
OPH
M4
Antares
SCO
M6
M20
M8
SGR
M17
M16
SCT
M11
ZENITH
Vega
M57
LYR
CYG
M29
VUL
M27
SGE
Altair
AQL
DEL
CAP
EQU
M15
Enif
Sadalmelik
AQR
Markab

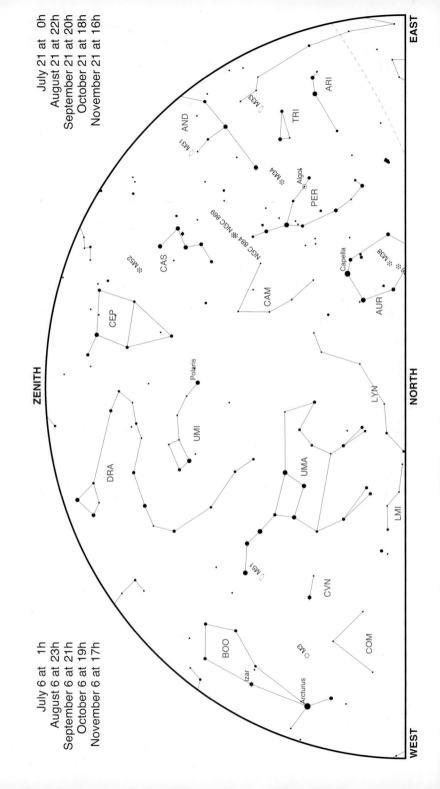

8N

July 21 at 0h
August 21 at 22h
September 21 at 20h
October 21 at 18h
November 21 at 16h

July 6 at 1h
August 6 at 23h
September 6 at 21h
October 6 at 19h
November 6 at 17h

EAST

ZENITH

NORTH

WEST

AND
M31
TRI
ARI
M33
M34
Algol
PER
NGC 884 NGC 869
M52
CAS
CAM
Capella
AUR
M38
M36
CEP
Polaris
UMI
LYN
DRA
UMA
LMI
M51
CVN
COM
BOO
Izar
Arcturus
M3

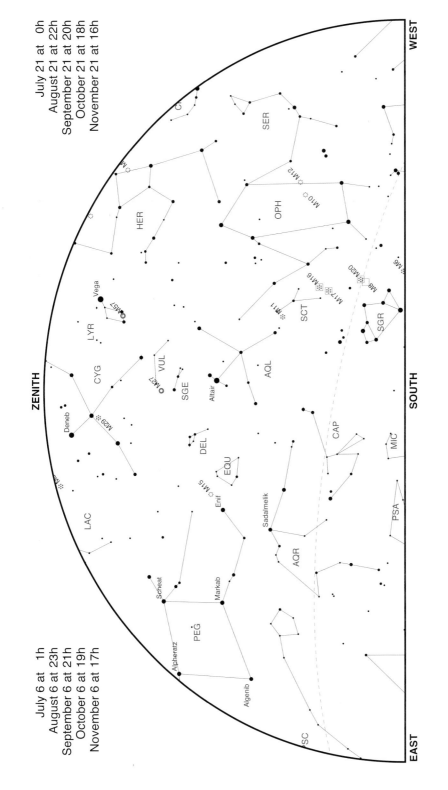

8S

WEST

EAST

SOUTH

ZENITH

July 21 at 0h
August 21 at 22h
September 21 at 20h
October 21 at 18h
November 21 at 16h

July 6 at 1h
August 6 at 23h
September 6 at 21h
October 6 at 19h
November 6 at 17h

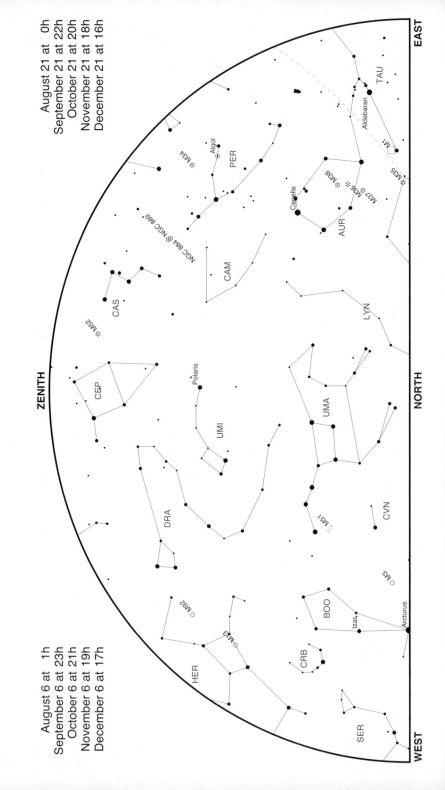

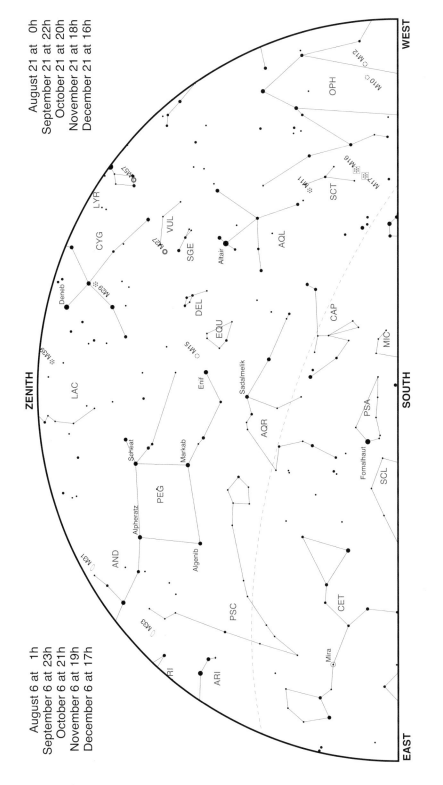

August 21 at 0h
September 21 at 22h
October 21 at 20h
November 21 at 18h
December 21 at 16h

August 6 at 1h
September 6 at 23h
October 6 at 21h
November 6 at 19h
December 6 at 17h

WEST

9S

ZENITH

EAST

SOUTH

OPH
M12
M10
M16
M17
M11
SCT
LYR
M57
CYG
VUL
M27
SGE
Altair
AQL
DEL
EQU
Deneb
M29
M15
Enif
CAP
M39
LAC
Sadalmelik
MIC
Scheat
PEG
Markab
AQR
PSA
Alpheratz
Fomalhaut
AND
Algenib
SCL
M31
PSC
M33
CET
ARI
ARI
Mira

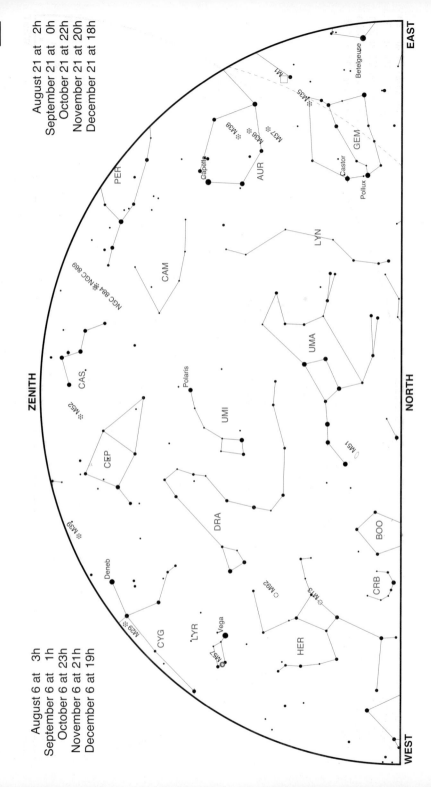

10N

August 6 at 3h
September 6 at 1h
October 6 at 23h
November 6 at 21h
December 6 at 19h

EAST

NORTH

WEST

ZENITH

Betelgeuse

GEM

Castor

Pollux

AUR

Capella

M38
M36
M37

M1

M35

PER

CAM

NGC 884 NGC 869

CAS.

M52

Polaris

UMI

UMA

LYN

M51

CEP

DRA

BOO

CRB

M39

M92

M13

HER

Deneb

CYG

LYR

Vega

M29

M57

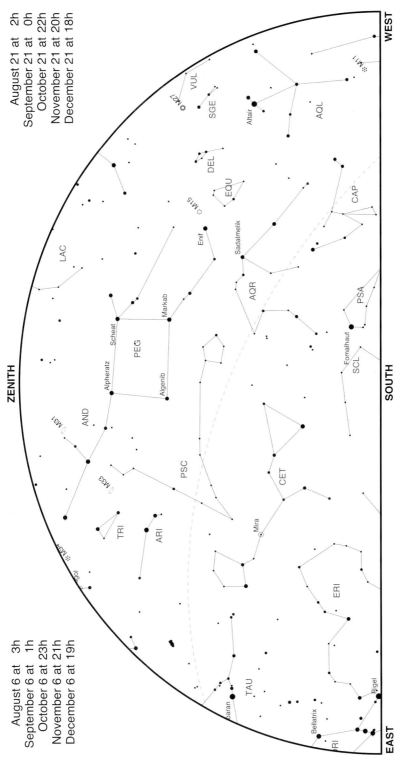

10S

WEST

VUL
M27
SGE
Altair
AQL
DEL
EQU
CAP
Enif
Sadalmelik
M15
LAC
AQR
PSA
Markab
Scheat
Alpheratz
PEG
Algenib
Fomalhaut
AND
SCL
M31
PSC
M33
CET
TRI
Mira
ARI
M34
Algol
ERI
TAU
Aldaran
Bellatrix
Rigel
RI

ZENITH

SOUTH

EAST

11N

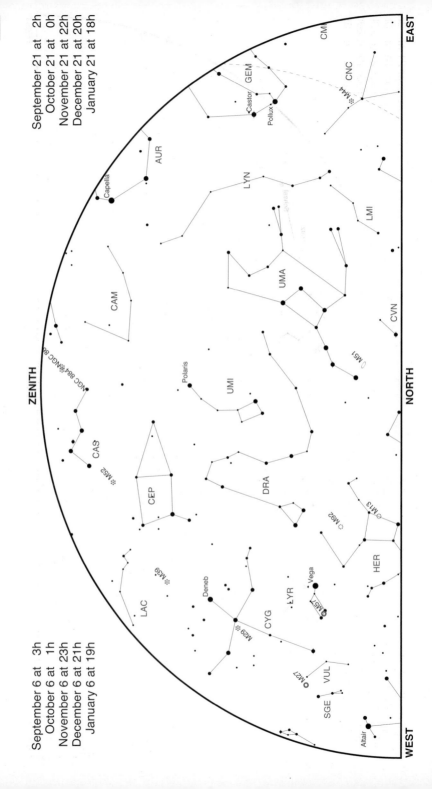

September 6 at 3h
October 6 at 1h
November 6 at 23h
December 6 at 21h
January 6 at 19h

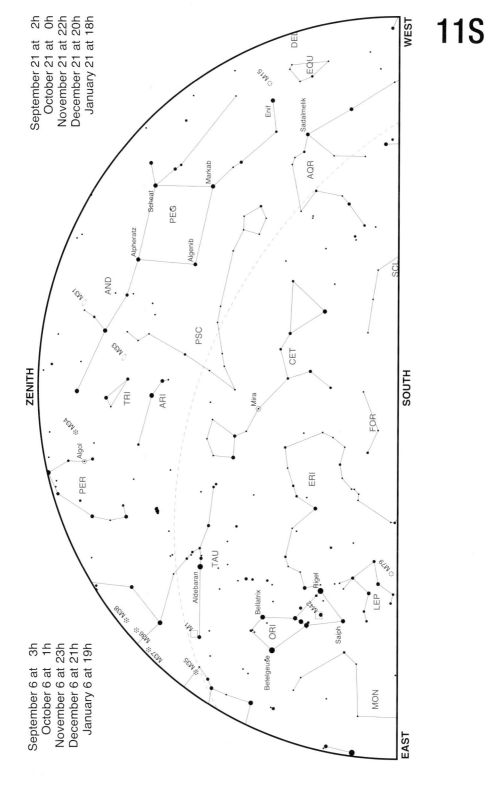

11S

September 21 at 2h
October 21 at 0h
November 21 at 22h
December 21 at 20h
January 21 at 18h

September 6 at 3h
October 6 at 1h
November 6 at 23h
December 6 at 21h
January 6 at 19h

WEST
EAST
SOUTH
ZENITH

DEL
EQU
M15
Enif
Sadalmelik
AQR
Markab
Scheat
PEG
Alpheratz
Algenib
AND
M31
PSC
M33
TRI
ARI
Mira
CET
FOR
SCL
Algol
M34
PER
ERI
M79
TAU
Aldebaran
M1
Rigel
Bellatrix
M42
LEP
Saiph
ORI
Betelgeuse
MON
M38
M36
M37
M35

12N

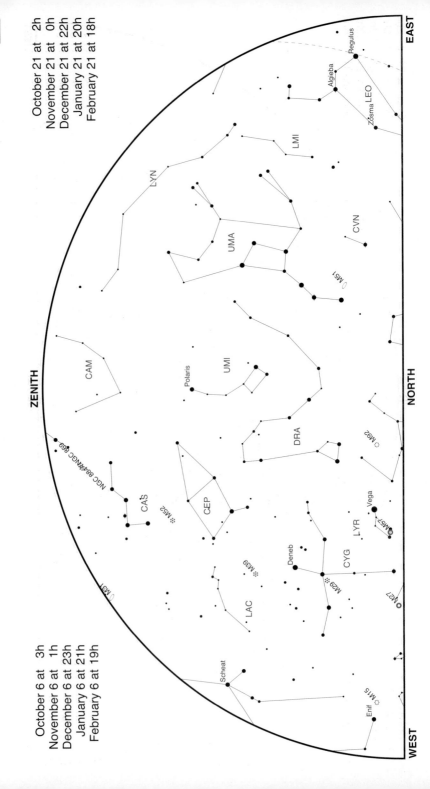

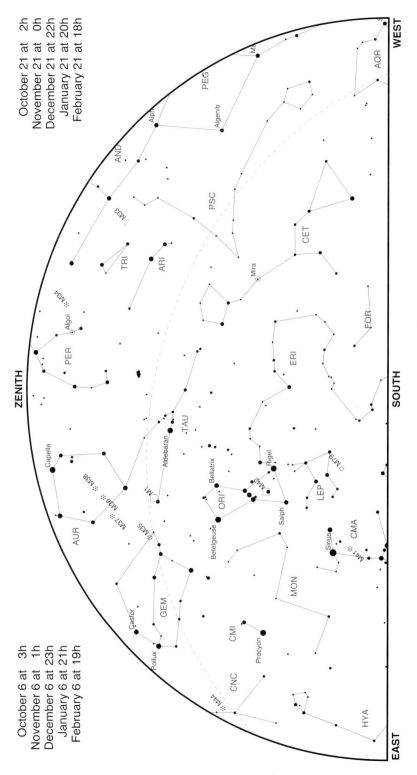

October 21 at 2h
November 21 at 0h
December 21 at 22h
January 21 at 20h
February 21 at 18h

October 6 at 3h
November 6 at 1h
December 6 at 23h
January 6 at 21h
February 6 at 19h

12S

WEST

EAST

SOUTH

ZENITH

AQR
PEG
Algenib
Alph
AND
M33
TRI
ARI
PSC
CET
Mira
FOR
ERI
PER
Algol
M34
Capella
M38
M36
M37
AUR
M35
M1
Aldebaran
TAU
Bellatrix
Rigel
M42
ORI
Saiph
Betelgeuse
M79
LEP
Sirius
CMA
M41
MON
Castor
Pollux
GEM
CMI
Procyon
CNC
M44
HYA

Southern Hemisphere Star Charts

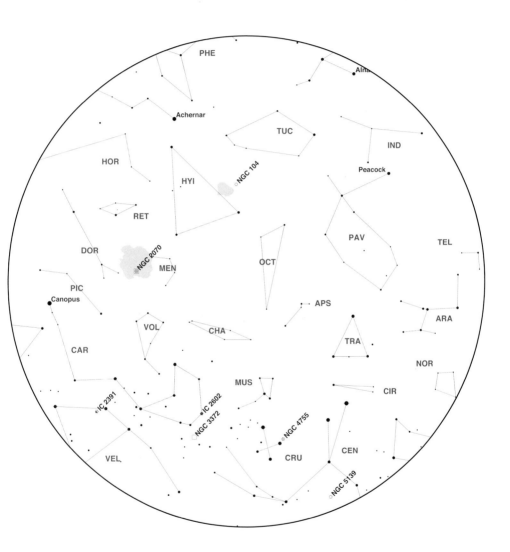

This chart shows stars lying at declinations between −45 and −90 degrees. These constellations are circumpolar for observers in Australia and New Zealand.

1N

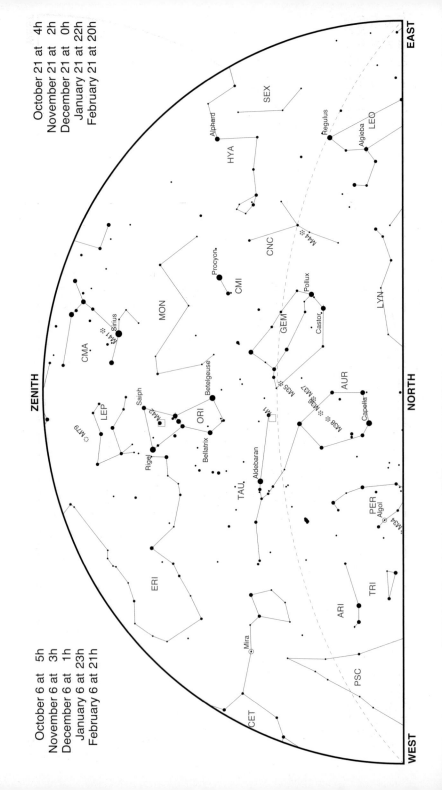

October 6 at 5h
November 6 at 3h
December 6 at 1h
January 6 at 23h
February 6 at 21h

EAST

ZENITH

NORTH

WEST

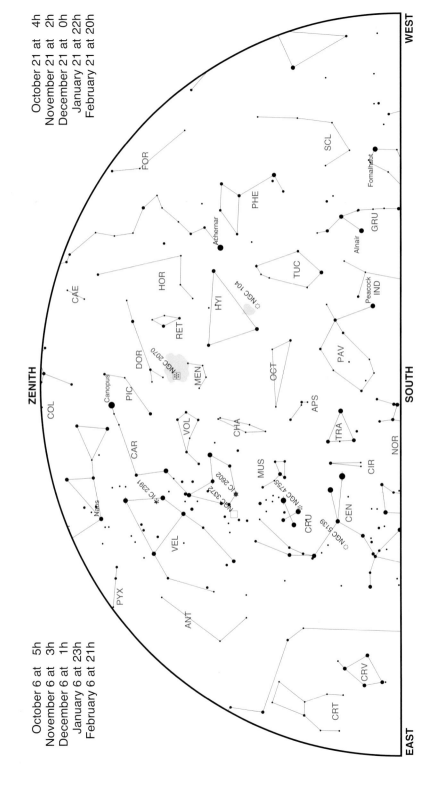

1S

WEST

EAST

ZENITH

SOUTH

NOR

October 21 at 4h
November 21 at 2h
December 21 at 0h
January 21 at 22h
February 21 at 20h

October 6 at 5h
November 6 at 3h
December 6 at 1h
January 6 at 23h
February 6 at 21h

FOR
SCL
PHE
Achernar
GRU
Alnair
Fomalhaut
TUC
NGC 104
Peacock
IND
HOR
HYI
CAE
RET
DOR
NGC 2070
MEN
OCT
PAV
PIC
Canopus
COL
CAR
VOL
CHA
APS
TRA
CIR
MUS
NGC 5139
CEN
NGC 4755
NGC 3372
CRU
IC 2602
IC 2391
NGC 2391
Naos
VEL
PYX
ANT
CRV
CRT

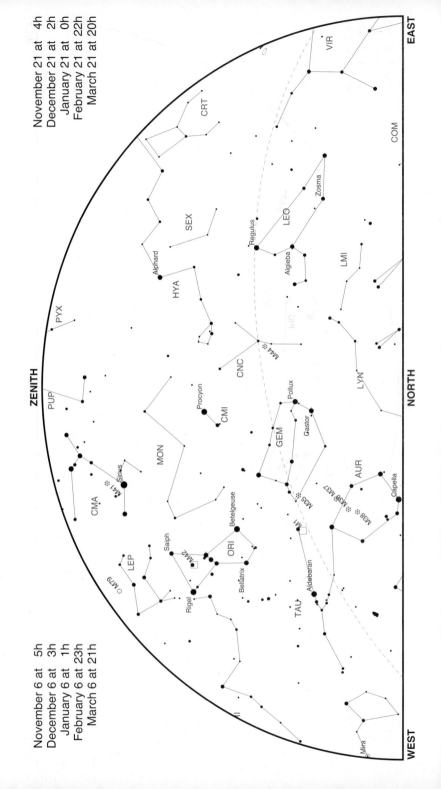

November 21 at 4h
December 21 at 2h
January 21 at 0h
February 21 at 22h
March 21 at 20h

November 6 at 5h
December 6 at 3h
January 6 at 1h
February 6 at 23h
March 6 at 21h

ZENITH

EAST

NORTH

WEST

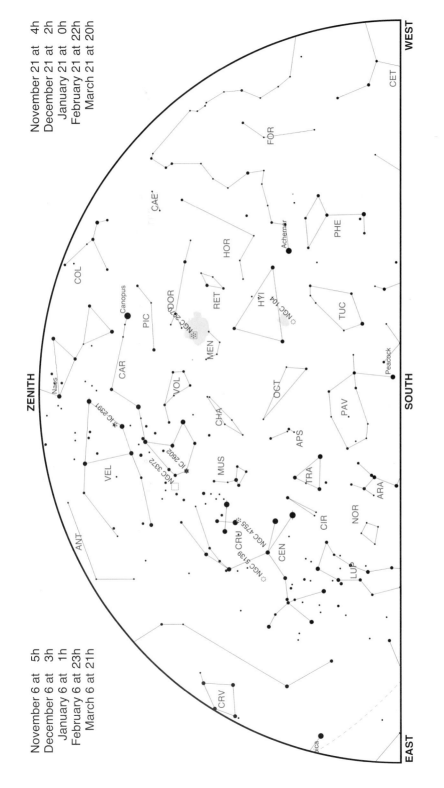

WEST

EAST

ZENITH

SOUTH

November 21 at 4h
December 21 at 2h
January 21 at 0h
February 21 at 22h
March 21 at 20h

November 6 at 5h
December 6 at 3h
January 6 at 1h
February 6 at 23h
March 6 at 21h

CET

FOR

CAE

HOR

PHE

Achernar

RET

HYI

NGC 104

TUC

MEN

NGC 2070

DOR

PIC

Canopus

COL

CAR

Naos

OCT

Peacock

SOUTH

PAV

VOL

CHA

APS

IC 2391

VEL

NGC 3372

IC 2602

MUS

TRA

ARA

ANT

NGC 4755

CRU

NGC 5139

CEN

CIR

NOR

LUP

CRV

Spica

3N

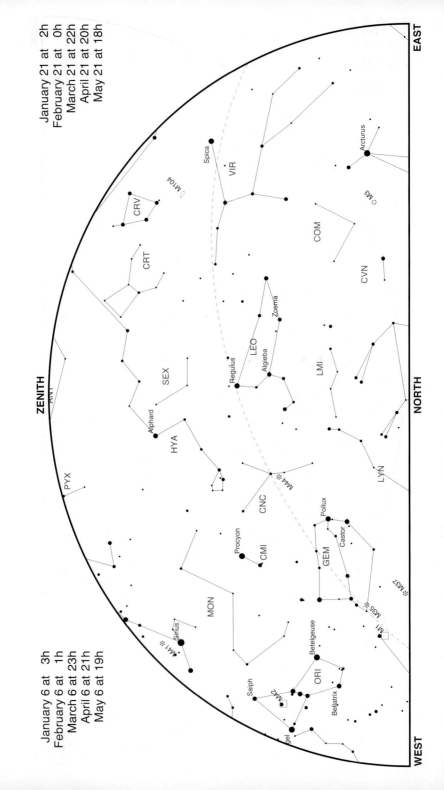

January 21 at 2h
February 21 at 0h
March 21 at 22h
April 21 at 20h
May 21 at 18h

January 6 at 3h
February 6 at 1h
March 6 at 23h
April 6 at 21h
May 6 at 19h

EAST

NORTH

WEST

ZENITH

Spica

VIR

Arcturus

M3

CRV

M104

CRT

COM

CVN

Zosma

LEO

Algieba

SEX

Regulus

LMI

Alphard

HYA

PYX

LYN

CNC

M44

CMI

Procyon

GEM

Pollux

Castor

MON

M37

M35

Sirius

M41

M1

ORI

Betelgeuse

M42

Saiph

Bellatrix

ANT

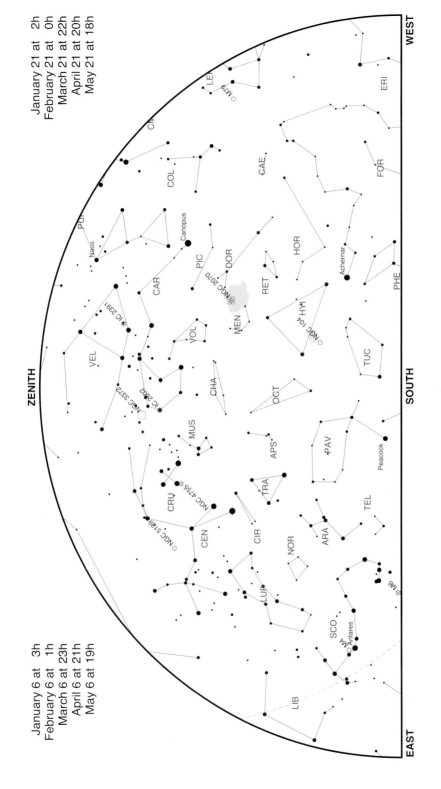

3S

January 21 at 2h
February 21 at 0h
March 21 at 22h
April 21 at 20h
May 21 at 18h

January 6 at 3h
February 6 at 1h
March 6 at 23h
April 6 at 21h
May 6 at 19h

WEST

EAST

SOUTH

ZENITH

LEP

ERI

CMA

COL

CAE

PUP

FOR

Naos

Canopus

PIC

DOR

HOR

PHE

Achernar

CAR

NGC 2070

RET

MEN

NGC 104

TUC

IC 2391

VOL

VEL

CHA

IC 2602

OCT

NGC 3372

PAV

MUS

APS

Peacock

CRU

NGC 4755

TRA

NGC 5139

CEN

CIR

TEL

ARA

NOR

LUP

SCO

M6

Antares

M4

LIB

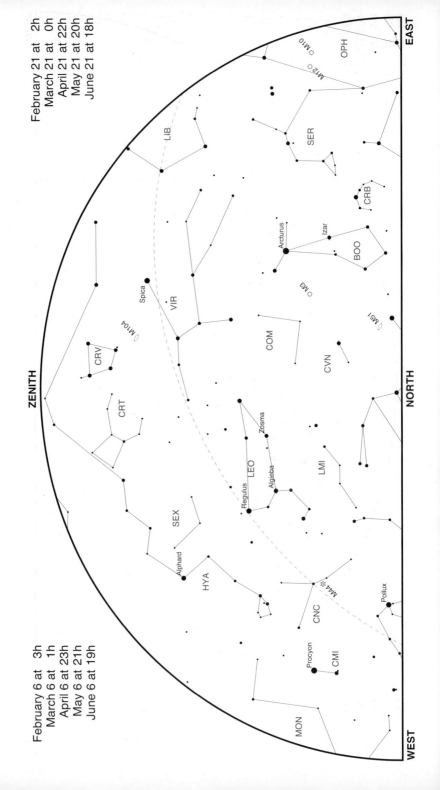

February 21 at 2h
March 21 at 0h
April 21 at 22h
May 21 at 20h
June 21 at 18h

February 6 at 3h
March 6 at 1h
April 6 at 23h
May 6 at 21h
June 6 at 19h

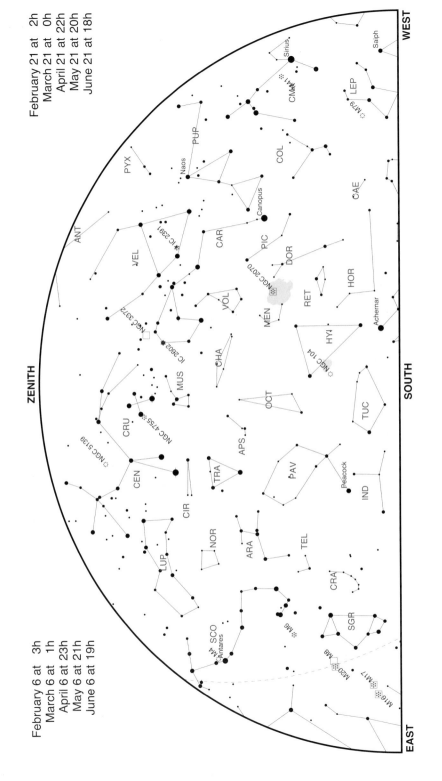

4S

WEST

EAST

SOUTH

ZENITH

February 21 at 2h
March 21 at 0h
April 21 at 22h
May 21 at 20h
June 21 at 18h

February 6 at 3h
March 6 at 1h
April 6 at 23h
May 6 at 21h
June 6 at 19h

5N

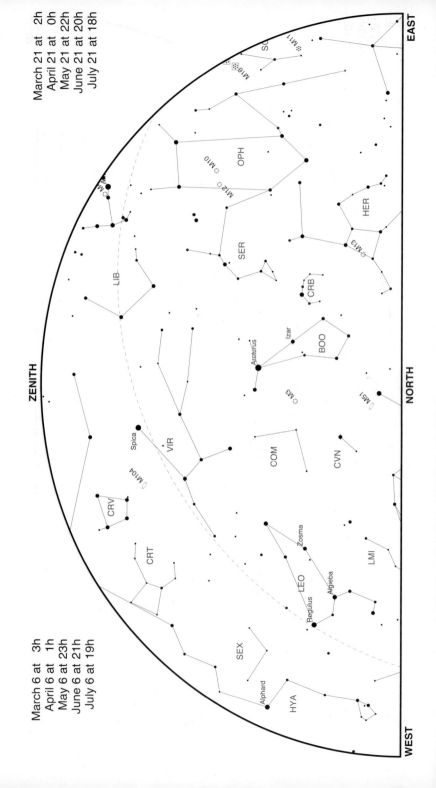

March 21 at 2h
April 21 at 0h
May 21 at 22h
June 21 at 20h
July 21 at 18h

EAST

ZENITH

NORTH

WEST

March 6 at 3h
April 6 at 1h
May 6 at 23h
June 6 at 21h
July 6 at 19h

SCO
M11
M16
M10
OPH
M12
HER
M13
SER
CRB
LIB
Izar
BOO
Arcturus
M51
M3
CVN
COM
VIR
Spica
M104
CRV
CRT
Zosma
LEO
Algieba
LMI
Regulus
SEX
Alphard
HYA

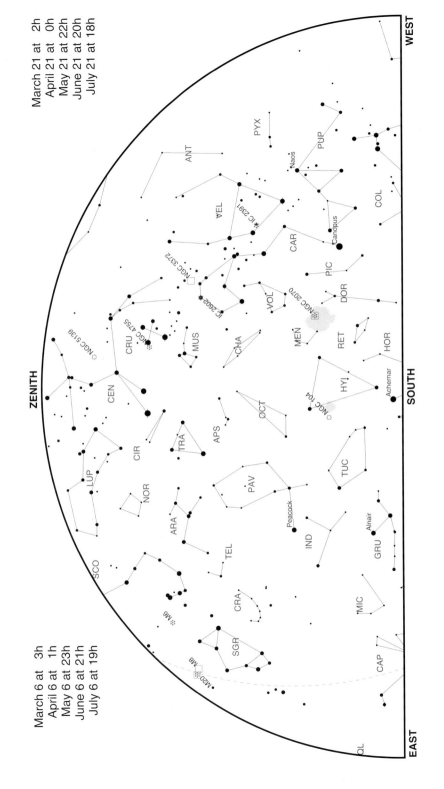

5S

WEST

ZENITH

EAST

SOUTH

March 21 at 2h
April 21 at 0h
May 21 at 22h
June 21 at 20h
July 21 at 18h

March 6 at 3h
April 6 at 1h
May 6 at 23h
June 6 at 21h
July 6 at 19h

PYX
ANT
PUP
Naos
VEL
IC 2391
CAR
Canopus
COL
PIC
NGC 3372
IC 2602
DOR
VOL
NGC 2070
CRU
NGC 4755
CHA
MEN
RET
HOR
NGC 5139
MUS
NGC 104
HYI
Achernar
CEN
OCT
TUC
CIR
APS
TRA
PAV
LUP
NOR
IND
Peacock
GRU
ARA
Alnair
SCO
TEL
MIC
M6
CRA
SGR
CAP
M20
M8

6N

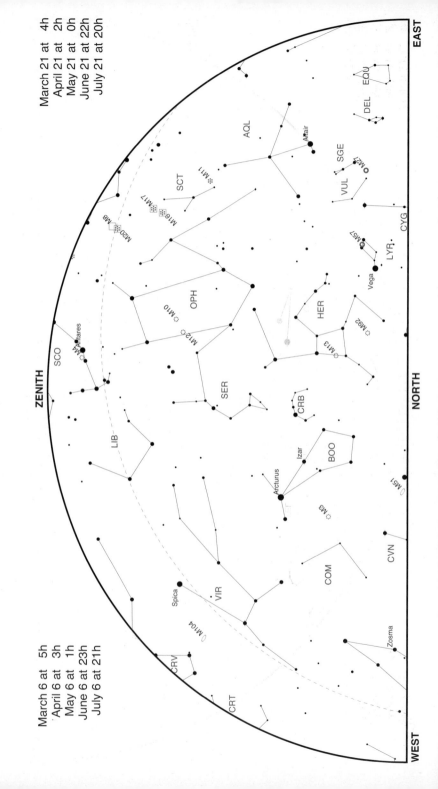

EAST

March 21 at 4h
April 21 at 2h
May 21 at 0h
June 21 at 22h
July 21 at 20h

March 6 at 5h
April 6 at 3h
May 6 at 1h
June 6 at 23h
July 6 at 21h

ZENITH

NORTH

WEST

EQU
DEL
AQL
Altair
SGE
M27
VUL
SCT
M11
M16 M17
M8
M20
LYR
Vega
M57
CYG
OPH
M10
HER
M92
SCO
Antares
M4
M12
M13
SER
CRB
LIB
Izar
BOO
Arcturus
M3
COM
CVN
M51
Spica
VIR
M104
Zosma
CRV
CRT

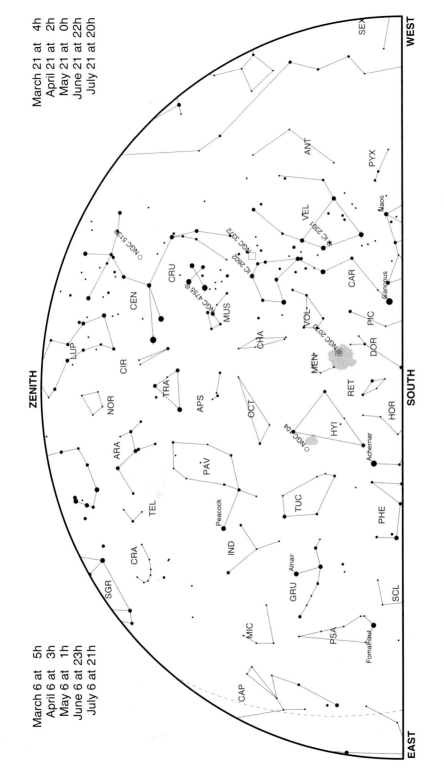

6S

WEST

EAST

ZENITH

SOUTH

March 21 at 4h
April 21 at 2h
May 21 at 0h
June 21 at 22h
July 21 at 20h

March 6 at 5h
April 6 at 3h
May 6 at 1h
June 6 at 23h
July 6 at 21h

SEX
ANT
PYX
VEL
CAR
PIC
DOR
RET
HOR
PHE
SCL
MEN
VOL
CHA
MUS
CRU
CEN
LUP
CIR
NOR
TRA
APS
OCT
HYI
TUC
PAV
ARA
TEL
IND
GRU
PSA
MIC
CRA
SGR
CAP

Naos
Canopus
Achernar
Peacock
Alnair
Fomalhaut

NGC 5139
NGC 3372
IC 2602
NGC 4755
IC 2391
NGC 2070
NGC 104

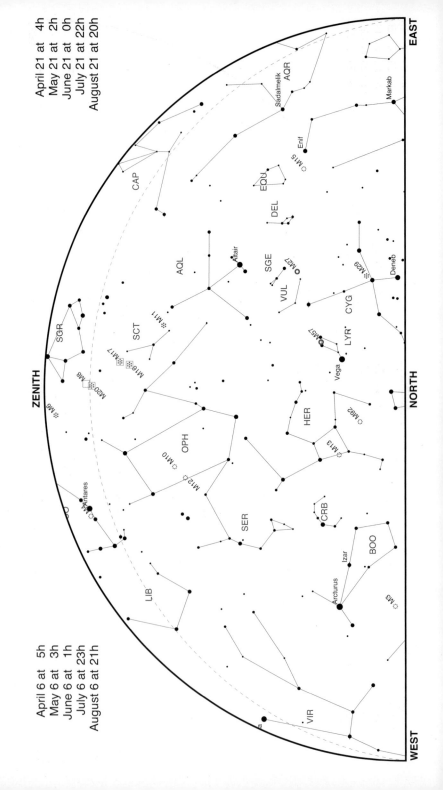

7N

April 6 at 5h
May 6 at 3h
June 6 at 1h
July 6 at 23h
August 6 at 21h

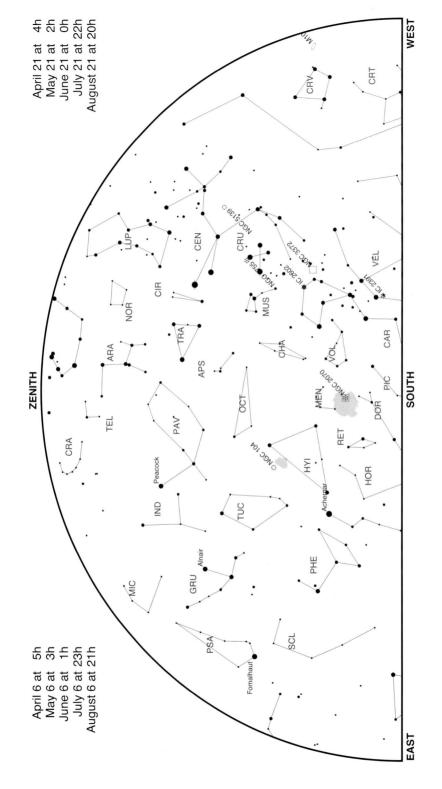

7S

WEST

SOUTH

EAST

ZENITH

April 21 at 4h
May 21 at 2h
June 21 at 0h
July 21 at 22h
August 21 at 20h

April 6 at 5h
May 6 at 3h
June 6 at 1h
July 6 at 23h
August 6 at 21h

CRV
CRT
M104
LUP
CEN
CRU
NGC 5139
NGC 3372
NGC 4755
NGC 2602
IC 2391
VEL
CIR
NOR
MUS
CHA
VOL
CAR
PIC
ARA
TRA
APS
MEN
NGC 2070
DOR
RET
HOR
TEL
OCT
HYI
Achernar
CRA
PAV
Peacock
IND
NGC 104
TUC
PHE
SCL
MIC
GRU
Alnair
PSA
Fomalhaut

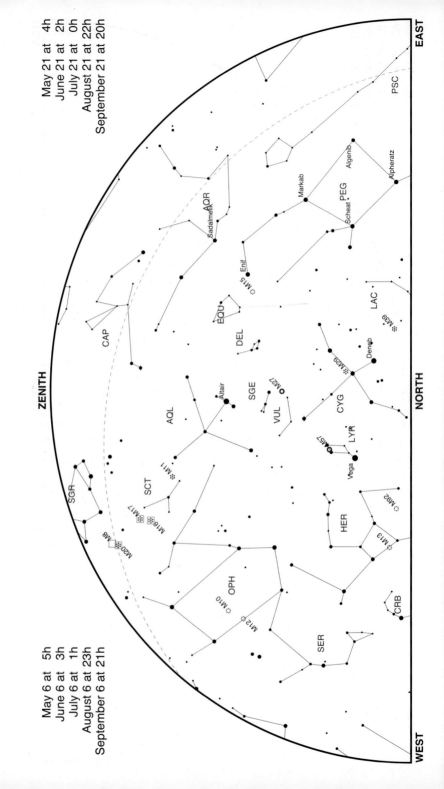

8N

May 21 at 4h
June 21 at 2h
July 21 at 0h
August 21 at 22h
September 21 at 20h

May 6 at 5h
June 6 at 3h
July 6 at 1h
August 6 at 23h
September 6 at 21h

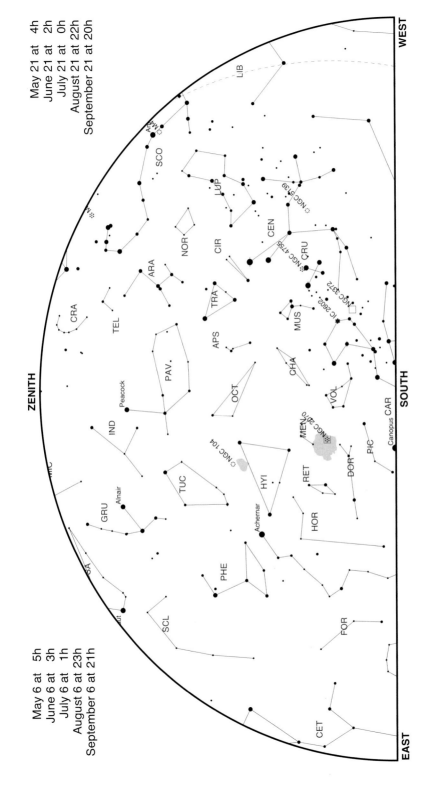

8S

WEST

ZENITH

SOUTH

EAST

May 21 at 4h
June 21 at 2h
July 21 at 0h
August 21 at 22h
September 21 at 20h

May 6 at 5h
June 6 at 3h
July 6 at 1h
August 6 at 23h
September 6 at 21h

LIB
SCO
M
LUP
NGC 5139
NOR
CEN
CIR
ARA
NGC 4755
CRU
CRA
TRA
NGC 4372
TEL
IC 2602
APS
MUS
PAV.
CHA
OCT
VOL
IND
Peacock
NGC 2070
MEN
DOR
CAR
Canopus
PIC
NGC 104
GRU
Alnair
HYI
RET
TUC
Achernar
HOR
PHE
SCL
FOR
CET

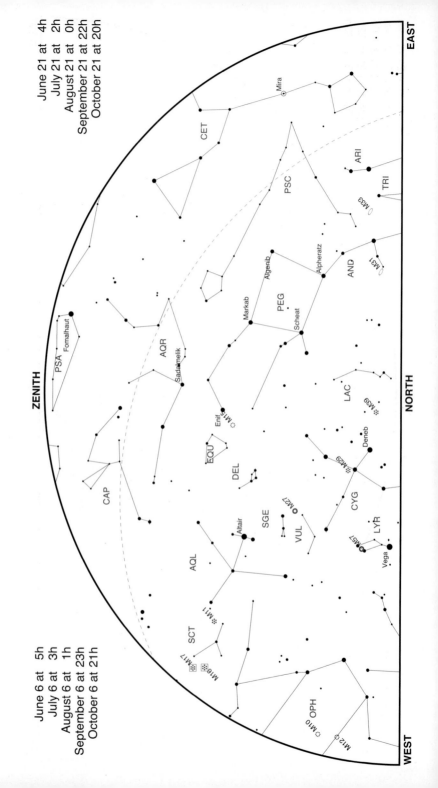

9N

June 21 at 4h
July 21 at 2h
August 21 at 0h
September 21 at 22h
October 21 at 20h

June 6 at 5h
July 6 at 3h
August 6 at 1h
September 6 at 23h
October 6 at 21h

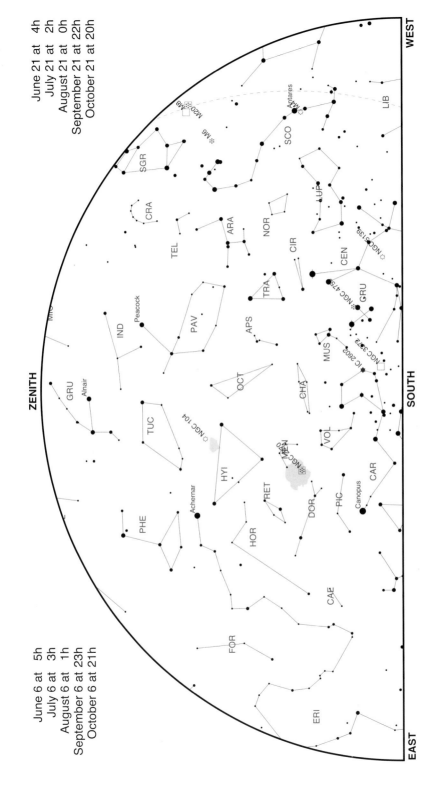

9S

June 21 at 4h
July 21 at 2h
August 21 at 0h
September 21 at 22h
October 21 at 20h

June 6 at 5h
July 6 at 3h
August 6 at 1h
September 6 at 23h
October 6 at 21h

WEST

EAST

ZENITH

SOUTH

10N

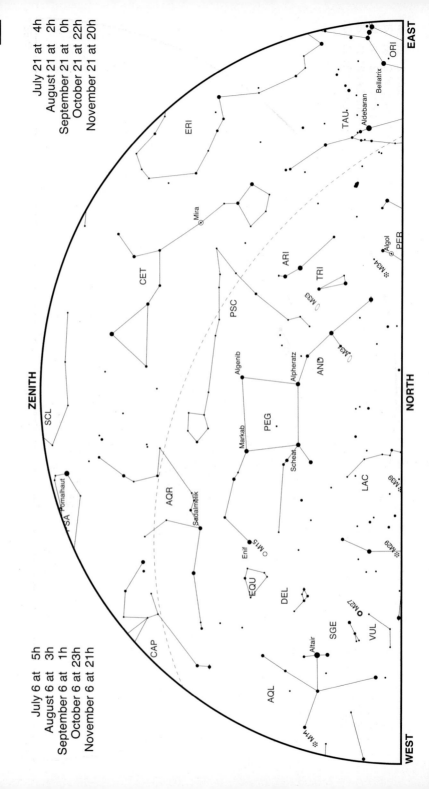

July 21 at 4h
August 21 at 2h
September 21 at 0h
October 21 at 22h
November 21 at 20h

July 6 at 5h
August 6 at 3h
September 6 at 1h
October 6 at 23h
November 6 at 21h

ZENITH

EAST

NORTH

WEST

ERI

TAU.

Aldebaran

Bellatrix

ORI

Mira

CET

ARI

TRI

M33

Algol

PER

M34

PSC

AND

M31

Algenib

Alpheratz

SCL

PEG

Markab

Scheat

LAC

M39

Fomalhaut

PSA

AQR

Sadalmelik

EQU

Enif

M15

DEL

M27

VUL

CAP

SGE

Altair

AQL

M11

M29

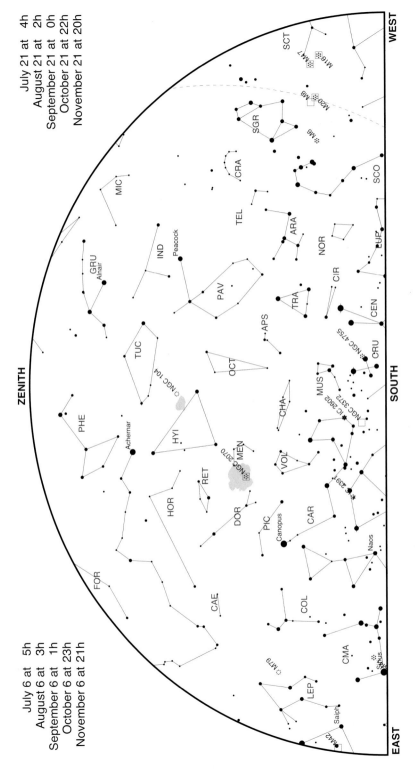

10S

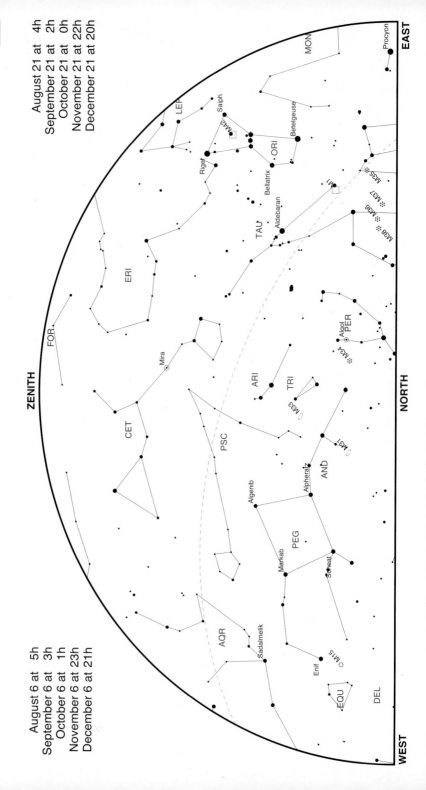

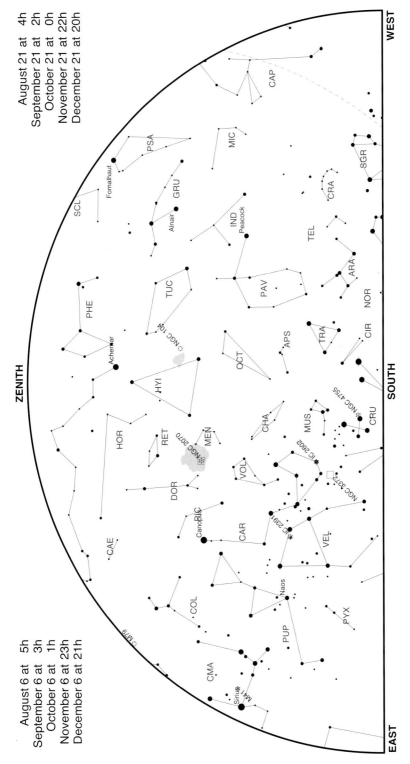

11S

WEST

ZENITH

SOUTH

EAST

CAP

PSA

MIC

SGR

Fomalhaut

GRU

CRA

SCL

Alnair

IND

Peacock

TEL

PHE

TUC

PAV

ARA

NOR

Achernar

NGC 104

HYI

OCT

APS

TRA

CIR

HOR

RET

MEN

CHA

MUS

NGC 4755

CRU

DOR

NGC 2070

VOL

IC 2602

NGC 3372

CAE

Canopus

CAR

NGC 2391

VEL

M79

COL

Naos

PUP

PYX

CMA

Sirius

M41

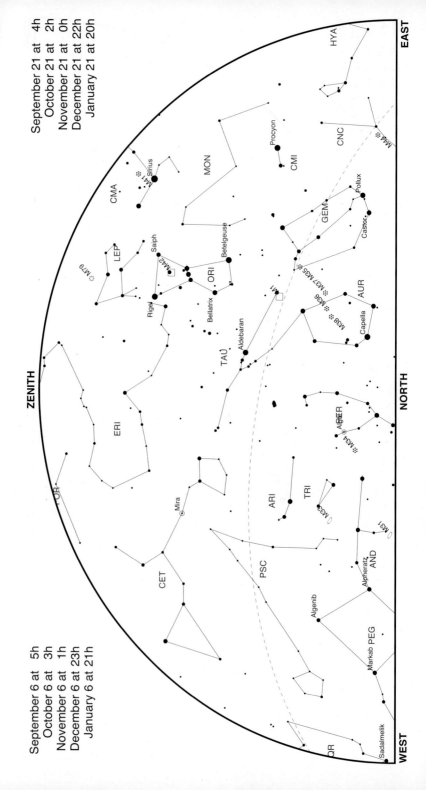

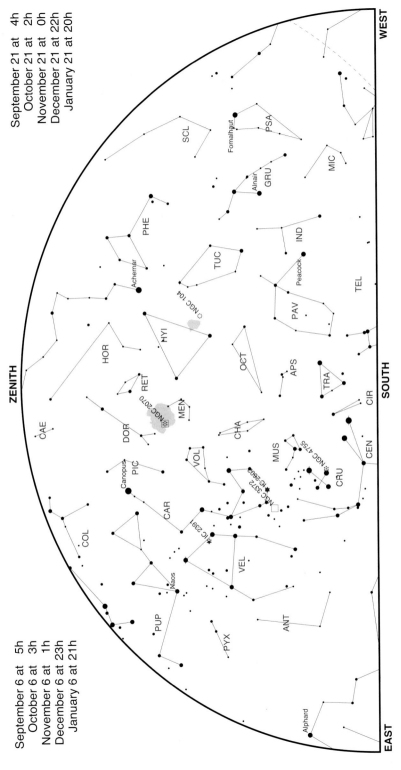

WEST

September 21 at 4h
October 21 at 2h
November 21 at 0h
December 21 at 22h
January 21 at 20h

SCL
PSA
Fomalhaut
MIC
GRU
Alnair
IND
PHE
Achernar
TUC
Peacock
NGC 104
HYI
PAV
TEL
HOR
OCT
APS
RET
MEN
TRA
DOR
NGC 2070
CHA
CIR
CAE
VOL
MUS
NGC 4755
CEN
ZENITH
PIC
CAR
NGC 2808
IC 2602
NGC 3372
CRU
Canopus
IC 2391
COL
VEL
Naos
PUP
ANT
PYX
Alphard

SOUTH

EAST

September 6 at 5h
October 6 at 3h
November 6 at 1h
December 6 at 23h
January 6 at 21h

The Planets in 2021

Lynne Marie Stockman

Mercury is one of the most elusive planets in the sky, never straying far from the Sun and thus never seen in truly dark skies. This year's best evening apparitions occur in April–June (northern temperate latitudes) and August–October (southern hemisphere). The optimal time to see Mercury in the east before sunrise is February–April for southern observers and October–November in the northern hemisphere. The August–October evening appearance of Mercury may not even be visible to planet watchers in the far north as the planet barely clears the horizon. Mercury has several close encounters with other planets but most of these bodies are near solar conjunction and difficult to see. The tiny planet is occulted by the Moon on 3 November and is also very close to the totally eclipsed Sun a month later on 4 December.

Venus is the morning star as the year begins, low in the east before sunrise and vanishing before superior conjunction in late March. It reappears as the evening star in April in what is an excellent apparition for observers in southern and equatorial latitudes. However, for astronomers in northern temperate regions, 2021 is the 'Year without Venus' as the planet struggles to reach much more than 15° in altitude. The brightest planet in the sky is occulted by the Moon in May and November, and passes close by three first-magnitude stars: Regulus (in July), Spica (in September) and Antares (in October). Venus reaches its maximum brightness in December.

Mars has a quiet year, progressing steadily through the constellations Pisces, Aries, Taurus, Gemini, Cancer, Leo, Virgo, Libra, and Scorpius before ending up in Ophiuchus. Mars has a long tenure as an evening sky object, beginning the year at magnitude −0.2 before fading to +1.8 by the time it reaches solar conjunction in early October. The red planet reappears in the east in November but remains embedded in the dawn twilight to the end of the year. Mars is occulted by the Moon three times this year, once in April and twice in December. It also passes close by Regulus (α Leonis) in July.

Jupiter begins the year in Capricornus, entering Aquarius in April before returning to Capricornus in August and then ending the year back in Aquarius. Jupiter is always very bright but it will be difficult to see during the first two months of the year as it undergoes solar conjunction in late January. It has several close encounters with Mercury and Venus around the time of conjunction but only the very close approach to Mercury in early March is likely to be easily visible. A morning sky object from February, it passes into the evening sky around June or July, eventually reaching opposition in August. The Moon keeps its distance this year, never approaching nearer than 3.3° but Jupiter does pass very close to fourth-magnitude star Theta (θ) Capricorni in February.

Saturn occupies the unremarkable constellation of Capricornus for the entire year. At conjunction with the Sun in late January, the ringed planet is not easily seen until toward the end of February when it appears in the dawn sky ahead of sunrise. It slowly pulls ahead of the Sun, reaching opposition in August. It is then visible in the evening sky until the end of the year. Saturn is approaching a ring plane crossing event in 2025 so the tilt of the rings is slowly decreasing. The rings are at their most open at the beginning of the year, closing down to a tilt of just under 17° in May before opening up slightly again. The Moon gets no closer than 3.2° to the planet this year; fourth-magnitude Theta (θ) Capricorni is the brightest star in close proximity (in May).

Uranus inhabits the dim constellation of Aries throughout 2021. It begins the year in the evening sky, vanishing in early April before reaching conjunction at the end of the month. It reappears in the east before dawn the following month, rising ever earlier and passing into the evening sky by July (northern hemisphere) or August (southern hemisphere). Opposition takes place in early November. The Moon makes a series of close passes by the planet, beginning over 3° away in January and coming to within 1.3° in October before moving away again. Sixth-magnitude Omicron (o) Arietis is the brightest star near the path of Uranus this year; the green ice giant passes by it mid-October whilst in retrograde.

Neptune spends the entirety of 2021 in the watery constellation of Aquarius. It is an evening sky object at the beginning of the year but vanishes in the glow of sunset by early to mid-February. Conjunction occurs in March and the eighth-magnitude planet reappears in the morning sky in April. It is best seen from southern latitudes during the middle part of the year and from the northern hemisphere at the beginning and end of the year. Opposition occurs in September. The most distant planet in the solar system does not pass by any bright stars this year and the Moon never gets closer than $4°$.

Phases of the Moon in 2021

Month	Last Quarter	New Moon	First Quarter	Full Moon
January	6	13	20	28
February	4	11	19	27
March	6	13	21	28
April	4	12	20	27
May	3	11	19	26
June	2	10	18	24
July	1 and 31	10	17	24
August	30	8	15	22
September	29	7	13	20
October	28	6	13	20
November	27	4	11	19
December	27	4	11	19

Eclipses in 2021

There are a minimum of four eclipses in any one calendar year, comprising two solar eclipses and two lunar eclipses. Most years have only four, as is the case with 2021, although it is possible to have five, six or even seven eclipses during the course of a year.

On 26 May there will be a total lunar eclipse which will be visible across the Pacific Ocean and regions of southern and eastern Asia, Japan, Australia and the western United States. The eclipse commences when the Moon enters the Earth's penumbra (partial shadow) at 08:48 UT and ends at 13:50 UT. The Moon begins to enter the Earth's umbra (full shadow) at 09:45 UT, with full eclipse lasting from 11:11 UT to 11:26 UT and maximum eclipse occurring at 11:20 UT. The Moon leaves the umbra at 12:52 UT.

An annular solar eclipse will take place on 10 June. The path of this eclipse will be restricted to the eastern tip of Russia, the Arctic Ocean, western Greenland and north eastern Canada. The partial eclipse will be observable from the north eastern regions of the United States, northern Europe and most of Russia. The eclipse commences at 08:12 UT and ends at 13:11 UT. Full eclipse will last from 09:50 UT to 11:34 UT with maximum eclipse at 10:42 UT.

There will be a partial lunar eclipse on 19 November which will be visible throughout most of eastern Russia, Japan, the Pacific Ocean, North America, Mexico, Central America, most of Australia and western South America. The eclipse commences at 06:02 UT and ends at 12:04 UT with the full eclipse lasting from 07:19 UT to 10:47 UT. Maximum eclipse occurs at 09:03 UT, when the Moon will be almost entirely immersed in the Earth's umbra, with only a thin sliver remaining outside this zone.

The path of totality of the total solar eclipse of 4 December will be confined to Antarctica and the southern Atlantic Ocean, although a small partial eclipse will be observable from most parts of South Africa. Observers located in Tasmania, and at various locations along the southern Australian coast, will see the partial phase as the sun sets. The eclipse begins at 05:29 UT and ends at 09:37 UT with total eclipse taking place between 07:00 UT and 08:06 UT and maximum eclipse at 07:33 UT.

Some Events in 2021

January	1	Saturn	Maximum ring opening (20.9°)
	2	Earth	Perihelion (0.9833 au)
	3/4	Earth	Quadrantid meteor shower (ZHR 120)
	24	Mercury	Greatest elongation east (18.6°)
	24	Saturn	Conjunction
	26	Uranus	East quadrature
	29	Jupiter	Conjunction

February	1	Mars	East quadrature
	3	Moon	Farthest perigee of the year (370,127 km)
	6	Venus, Saturn	Planetary conjunction (0.4° apart, 12° from the Sun)
	8	Mercury	Inferior conjunction

March	4	4 Vesta	Opposition in Leo (magnitude +5.9)
	5	Mercury, Jupiter	Planetary conjunction (0.3° apart, 27° from the Sun)
	6	Mercury	Greatest elongation west (27.3°)
	11	Neptune	Conjunction
	14	Venus, Neptune	Planetary conjunction (0.4° apart, 3° from the Sun)
	20	Earth	Equinox
	26	Venus	Superior conjunction

April	17	Mars	Lunar occultation (59° from the Sun)
	19	Mercury	Superior conjunction
	22/23	Earth	Lyrid meteor shower (ZHR 18)
	23	Venus, Uranus	Planetary conjunction (0.2° apart, 7° from the Sun)
	24	Mercury, Uranus	Planetary conjunction (0.7° apart, 6° from the Sun)
	30	Uranus	Conjunction

May	3	Saturn	West quadrature
	6/7	Earth	Eta (η) Aquariid meteor shower (ZHR 30)
	11	Moon	Farthest apogee of the year (406,512 km)
	12/13	Venus	Lunar occultation (12° from the Sun)
	17	Mercury	Greatest elongation east (22.0°)
	20	Saturn	Minimum ring opening (16.7°)
	21	Jupiter	West quadrature
	26	Earth, Moon	Total lunar eclipse
	26	Moon	Super Moon (largest apparent diameter of the year)
	29	Mercury, Venus	Planetary conjunction (0.4° apart, 17° from the Sun)

	6	3 Juno	Opposition in Ophiuchus (magnitude +10.1)
	9	Earth	Tau (τ) Herculid meteor shower (ZHR low)
	10	Earth, Moon	Annular solar eclipse
June	11	Mercury	Inferior conjunction
	13	Neptune	West quadrature
	21	Earth	Solstice
	23	Mars	0.3° south of Praesepe (M44)

	3	Venus	0.1° north of Praesepe (M44)
	4	Mercury	Greatest elongation west (21.6°)
	5	Earth	Aphelion (1.107 au)
July	13	Venus, Mars	Planetary conjunction (0.5° apart, 28° from the Sun)
	17	134340 Pluto	Opposition in Sagittarius (magnitude +14.6)
	21	Venus	1.0° north of Regulus (α Leonis)
	28/29	Earth	Delta (δ) Aquariid meteor shower (ZHR 20)
	29	Mars	0.6° north of Regulus (α Leonis)

	1	Mercury	Superior conjunction
	2	Saturn	Opposition in Capricornus (magnitude +0.2)
	7	Uranus	West quadrature
August	12/13	Earth	Perseid meteor shower (ZHR 80)
	19	Mercury, Mars	Planetary conjunction (0.1° apart, 16° from the Sun)
	19	Jupiter	Opposition in Aquarius (magnitude −2.9)
	22	Moon	Seasonal Blue Moon
	30	Moon	Nearest apogee of the year (404,100 km)

	11	2 Pallas	Opposition in Pisces (magnitude +8.5)
	14	Mercury	Greatest elongation east (26.8°)
September	14	Neptune	Opposition in Aquarius (magnitude +7.8)
	20	Moon	Harvest Moon
	22	Earth	Equinox

	7/8	Earth	Draconid meteor shower (ZHR 10)
	8	Mars	Conjunction
	9	Mercury	Inferior conjunction
October	10	Earth	Southern Taurid meteor shower (ZHR 5)
	21/22	Earth	Orionid meteor shower (ZHR 25)
	25	Mercury	Greatest elongatiion west (18.4°)
	29	Venus	Greatest elongation east (47.0°)
	30	Saturn	East quadrature

November	3	Mercury	Lunar occultation (15° from the Sun)
	5	Uranus	Opposition in Aries (magnitude +5.7)
	8	Venus	Lunar occultation (47° from the Sun)
	10	Mercury, Mars	Planetary conjunction (1.0° apart, 11° from the Sun)
	12	Earth	Northern Taurid meteor shower (ZHR 5)
	15	Jupiter	East quadrature
	17/18	Earth	Leonid meteor shower (ZHR varies)
	19	Earth, Moon	Partial lunar eclipse
	27	1 Ceres	Opposition in Taurus (magnitude +7.0)
	29	Mercury	Superior conjunction

December	2/3	Mars	Lunar occultation (18° from the Sun)
	4	Earth, Moon	Total solar eclipse
	4	Moon	Nearest perigee of the year (356,794 km)
	4	Moon, Mercury	Planetary conjunction (0.02° apart, 3° from the Sun)
	12	Neptune	East quadrature
	13/14	Earth	Geminid meteor shower (ZHR 75+)
	19	Moon	Micro Moon (smallest apparent diameter of the year)
	21	Earth	Solstice
	21/22	Earth	Ursid meteor shower (ZHR 10)
	31	Mars	Lunar occultation (27° from the Sun)

The entries for meteor showers state the date of peak shower activity (maximum). The figure quoted in parentheses in column 4 alongside each meteor shower entry is the expected Zenith Hourly Rate (ZHR) for that particular shower at maximum. For a more detailed explanation of ZHR, and for further details of the individual meteor showers listed here, please refer to the article *Meteor Showers in 2021* located elsewhere in this volume. For more on each of the eclipses occurring during the year, please refer to the information given in *Eclipses in 2021*.

Monthly Sky Notes and Articles

Morning Apparition of Venus
June 2020 to March 2021

Legend:
— 52° North
— 35° South

Chart markers:
- 1 Jan
- 1 Feb
- 1 Dec
- 1 Nov
- 1 Oct
- 1 Sep
- 1 Aug
- 13 Aug (GE west)
- 1 Jul
- 1 Mar

Axis labels: 0°, 10°, 20°, 30°, 40°, 50°

Horizon directions: NE, E, SE

January

New Moon: 13 January
Full Moon: 28 January

MERCURY is visible in the west soon after sunset, with northern vantage points favoured over the southern hemisphere. Its brightness spans two magnitudes, beginning the month at −1.0 and ending at +1.1. The tiny planet is 1.6° south of Saturn on 10 January and 1.4° south of Jupiter the following day but both events will be tricky to observe as all three planets are only about 13° away from the Sun at the time. The nearly New Moon passes 2.3° south of Mercury on 14 January; under perfect seeing conditions sharp-eyed planet watchers may spot it. Mercury reaches greatest elongation east (18.6°) on 24 January and enters into retrograde motion on the penultimate day of the month.

Evening Apparition of Mercury
20 December to 8 February

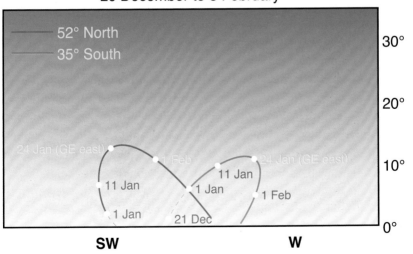

VENUS is the morning star, best viewed from the tropics. Currently at magnitude −3.9, it is only 20° or less above the eastern horizon, and appears a little lower in the sky every morning. A telescopic view reveals a small waxing gibbous disk. The waning crescent Moon pairs up with Venus on 11 January, with the two objects just 1.5° apart.

EARTH reaches the point in its orbit where it is nearest to the Sun on the second day of January. This year's perihelion distance is 0.9833 au or 147,100,000 km. The Quadrantid meteor shower peaks soon after. For more details see *Meteor Showers in 2021* elsewhere in this volume. The Moon passes 2.3° north of the open star cluster Praesepe (M44) twice this month, on 1 January and again on 28 January. However, the brightness of the full or nearly full Moon makes spotting the star cluster difficult.

MARS is a zero-magnitude object as 2021 gets underway. Briefly in Pisces, Mars moves into the neighbouring constellation of Aries on 5 January. It passes 1.7° north of Uranus on 20 January but a telescope may be necessary to see the fainter planet because of the light from the First Quarter Moon. Mars watchers in northern temperate latitudes have the best views of the red planet this month, with Mars not setting until after midnight. *A finder chart showing the position of Mars throughout the first six months of 2021 can be found in the February sky notes. Background stars are shown to magnitude 5.5*

JUPITER and Saturn underwent a 'Great Conjunction' last month but now Jupiter is leaving Saturn behind in the constellation of Capricornus. It arrives at solar conjunction on 29 January, making it difficult to see low in the west at sunset at the beginning of the year. The largest planet in the solar system is at its maximum declination south for year on 1 January and is found 1.5° away from Mercury ten days later, but with the two planets only 13° from the Sun, it is possible that this event will go unnoticed. *A finder chart showing the position of Jupiter (and Saturn) throughout 2021 can be found in the August sky notes. Background stars are shown to magnitude 5.5*

SATURN begins the year low in the west, setting during evening twilight and vanishing from view long before its conjunction with the Sun on 24 January.

Like Jupiter, it is at its most southerly declination of the year at the outset of 2021 which is when it also exhibits its maximum ring tilt of 20.9°. Saturn and Mercury have a close encounter on 10 January but the two objects are only 12° away from the Sun at the time, rendering the planetary conjunction very difficult to see. *A finder chart showing the position of Saturn (and Jupiter) throughout 2021 can be found in the August sky notes. Background stars are shown to magnitude 5.5*

URANUS is present during the evening and early morning hours, and is best seen from northern latitudes this month. Found in the constellation of Aries, Uranus reaches declination +13.2°, as far south as it gets in 2021, on 12 January and two days later, ends its retrograde track across the sky to resume direct motion. Mars is less than 2° north of Uranus on 20 January but the First Quarter Moon may make it difficult to spot sixth-magnitude Uranus with the naked eye. The green ice giant marks the 35th anniversary of the historic **Voyager 2** flyby on 24 January and reaches east quadrature two days later. *A finder chart showing the position of Uranus throughout 2021 can be found in the November sky notes. Background stars are shown to magnitude 8.*

NEPTUNE is above the horizon in the evening hours at the beginning of 2021. It is at its maximum southerly declination for the year on the first day of January and is slowly moving eastwards away from fourth-magnitude star Phi (φ) Aquarii, passing 0.4° south of the sixth-magnitude star 96 Aquarii on 21 January. Neptune itself is only magnitude +7.9 so a small telescope and preferably a moonless evening are needed to see it. *A finder chart showing the position of Neptune throughout 2021 can be found in the September sky notes. Background stars are shown to magnitude 10.*

Astronomical Illustrations of the Nuremberg Chronicle

Richard H. Sanderson

Liber Chronicarum (Book of Chronicles), a masterpiece of Renaissance printing, was the most lavishly-illustrated book of its time. More commonly known as the *Nuremberg Chronicle* after the city in Germany where it was created and published, this folio-sized work contains fascinating woodcuts that represent celestial objects and events.

The *Nuremberg Chronicle* is a history of the world that features accounts of biblical and historical events, including natural phenomena such as earthquakes, eclipses and comets. It was written by Hartmann Schedel, a medical doctor and avid book collector whose library included works on astronomy. Schedel drew his information from the Bible and numerous early chronicles.

Eclipses are represented in the *Nuremberg Chronicle* by a woodcut depicting the sun and moon. (Richard H. Sanderson / Author's Collection)

Nuremberg artist and printmaker Michael Wolgemut and his stepson, Wilhelm Pleydenwurff, spent two to three years creating roughly 645 woodcut illustrations for the book. Wolgemut was a mentor to the renowned artist Albrecht Dürer. The Chronicle contains more than 1,800 illustrations, with many of the smaller images appearing more than once. Some original owners paid more to have the pictures in their copies hand coloured. Many of the illustrations in the Chronicle are captivating and beautiful, especially the 26 spectacular, double-page city views.

The book was published by one of the fifteenth-century's most accomplished printers, Anton Koberger. Publication of the Latin first edition, with an

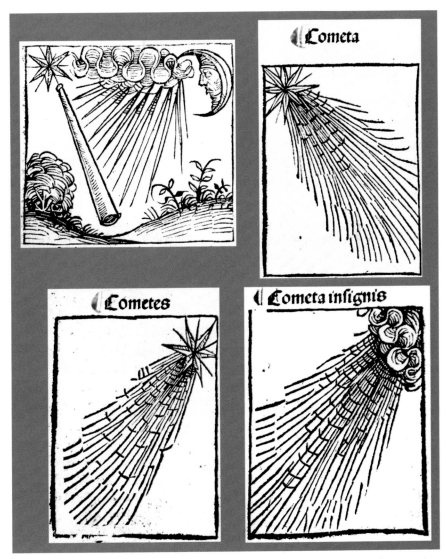

Four different woodcuts were used to create 13 generic comet illustrations in the *Nuremberg Chronicle*. (Richard H. Sanderson/Author's Collection)

estimated print run of around 1,400 copies, was completed in 1493. A vernacular German-language edition appeared later that year. That makes the *Nuremberg Chronicle* an "incunabulum," a book printed in Europe prior to 1501, which is an important distinction among book connoisseurs.

I was intrigued by a woodcut from the *Nuremberg Chronicle* that was reproduced in several comet books published during the 1985-1986 visit of Halley's Comet. Some captions identified it as a drawing of Halley's Comet made in 684 or as the oldest known illustration of the famous comet, but as we shall see, these descriptions are not completely accurate.

A few years ago, I acquired seven leaves from the Latin first edition. Reverently touching them for the first time, I was amazed by the supple, high-quality rag paper and the bold black ink. The pages of the *Nuremberg Chronicle* were printed on both sides, but only the fronts are numbered. When referring to folio (page) numbers, the letters "r" and "v" are added to the Roman numerals to differentiate between recto (front) and verso (back).

Among the images that adorn the *Nuremberg Chronicle* is a striking woodcut on folio CLVIIr (the front side of page 157) showing a sombre-faced sun and moon. This sun-moon image is a generic representation of solar and lunar eclipses which, according to the text, were followed by a severe pestilence. The same eclipse woodcut appears on folio LXXVIr. Many of the astronomical events recorded in the *Nuremberg Chronicle* are regarded as omens.

An unusual pair of woodcuts on folio CCIIIv depicts a triple moon and triple sun. The accompanying text describes how three suns appeared in the west, with the middle one setting after the other two had disappeared. It also mentions a similar display involving the moon. These are twelfth-century sightings of parhelia (sun dogs) and paraselenae (moon dogs), two relatively common atmospheric phenomena.

The Ensisheim meteorite fall occurred during the production of the book and is included toward the end. A striking woodcut on folio CCLVIIr depicts a rock, greatly exaggerated in size, emerging from a cloud above the town and surrounded by shafts of light. The account tells us that a large stone "fell from the sky at Ensisheim" at noon on 7 November 1492.

The 13 comet images found in the *Nuremberg Chronicle* were created using four different woodcuts. They appear in sections of the book covering the fifth to the fifteenth centuries, although it is not possible to connect all of these pictures to specific comets. One woodcut is linked to two different comets. Clearly, the images themselves are not observations or likenesses of actual comets from those time periods. Instead, they must be regarded as fifteenth-century stylized illustrations that are meant to embellish the textual accounts. Each of the seven leaves I purchased features a comet image and I managed to obtain exemplars of all four comet woodcuts.

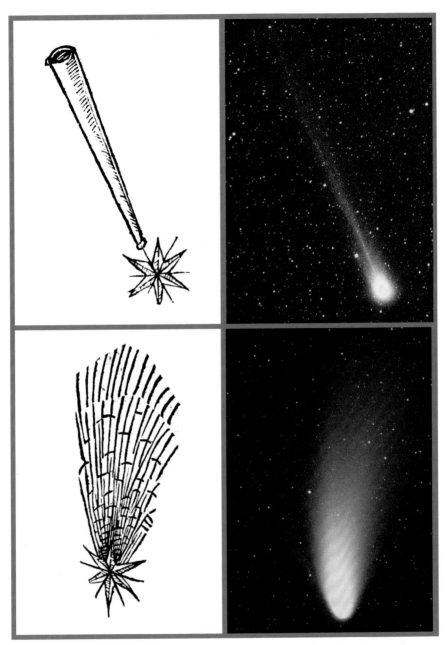

The comet woodcuts in the *Nuremberg Chronicle* are moderately realistic depictions, as shown in this comparison with Comet Hyakutake (top) and Comet Hale-Bopp. The woodcuts have been inverted so that the tails point upward. (Richard H. Sanderson/Author's Collection)

In each of the Chronicle's comet illustrations, a star is positioned at the head of the comet, perhaps representing a bright coma, followed by a tail. Attempting to depict nebulous celestial objects using woodcuts must have been a challenging endeavour. The comets are also shown up-side-down in the book but, with one exception, they are easily recognizable. Perhaps Michael Wolgemut had seen the bright comet that decorated the heavens in early 1472 and is mentioned in conjunction with the book's final comet woodcut.

The exception is the first comet image, which appears on folio CXLIv. Of the four woodcut variants used to represent comets, this one appears once while the others are used multiple times. Unlike the others, it conflates the comet, which oddly resembles a spyglass with a star at one end, with other events that were collectively seen as "signs" in the sky. The corresponding text tells us about fiery appearances in the north (aurora borealis?), a darkened moon (lunar eclipse) and lightning, as well as a comet that appeared at Toulouse. The long, narrow tail shown in the woodcut reminds me of the naked-eye ion tail displayed by Comet Hyakutake. The appendages depicted in the other comet woodcuts resemble dust tails.

Folio CLVIIr features both the iconic eclipse image already discussed and the well-publicized comet woodcut linked to the 684 visit of Halley's Comet. My copy of this page came from a "rubricated" version of the Chronicle, which means that the original purchaser paid extra to have a scribe add red letters to highlight the beginnings of sections. The red ink is still vibrant after more than 600 years!

Does the illustration on folio CLVIIr really refer to Halley's Comet? Possibly, but a definitive connection cannot be made. The image appears in a section of the book spanning 684, the year of Halley's return, and 693. The undated description states that a comet appeared for "three months continuously," hinting at a splendid cosmic display. A woodcut included later in the book can definitely be linked to the appearances of Halley's Comet in 1456 and another comet in 1457. The Chronicle's accuracy improves as the dates of historical events approach the time of its publication.

Thanks to its spectacular graphic artistry, the *Nuremberg Chronicle* was very popular in its day and it remains an electrifying example of the art of the book. The astronomical woodcuts it contains are fifteenth-century generic representations of celestial events rather than first-hand observations, but in no way does that diminish their tantalizing allure.

Sexta etas mundi

Foliū CLVII

Folio CLVIIr includes a comet woodcut that might link to the 684 visit of Halley's Comet. (Richard H. Sanderson/Author's Collection)

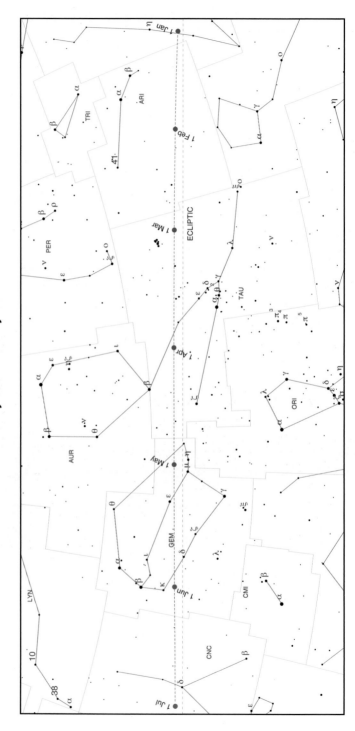

Mars in the Evening Sky
January 2021 to July 2021

February

New Moon: 11 February
Full Moon: 27 February

MERCURY is still an evening sky object at the beginning of February but soon vanishes below the horizon as it undergoes inferior conjunction on 8 February. It quickly passes into the morning sky in what is the best dawn apparition for southern and equatorial latitudes. Morning appearances of Mercury are characterised by a rapid brightening of the planet as it speeds away from the Sun; in this instance Mercury reaches magnitude +0.3 by the end of the month. The nearest planet to the Sun passes by Venus on 13 February and Jupiter the day after but never gets closer than about 4° to either of them. Mercury returns to direct motion on 20 February and continues to climb higher above the eastern horizon.

Morning Apparition of Mercury
8 February to 19 April

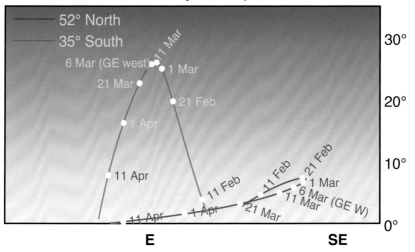

VENUS is visible low in the east at dawn. It is drawing nearer to the Sun every day as it approaches superior conjunction late next month. Observers in northern temperate latitudes will be hard-pressed to see the magnitude −3.9 planet before sunrise, with the best views reserved for early risers in the southern hemisphere. The morning star is only 0.4° south of Saturn on 6 February and is a similar distance from Jupiter five days later. However, the altitude of all three objects will be quite low, making it difficult to observe these planetary conjunctions.

EARTH experiences the most distant lunar perigee of the year on 3 February. On 25 February, the waxing gibbous Moon passes 2.4° north of Praesepe (M44), the famous open star cluster in the constellation of Cancer.

MARS is getting fainter with every passing night, beginning the month at magnitude +0.4 and ending at +0.9. An evening sky object best seen from the northern hemisphere, Mars is at east quadrature on the first day of the month and presents a distinctly gibbous appearance in a telescope. It occupies the constellation Aries until 23 February when it moves into Taurus.

JUPITER, shining at magnitude −2.0, and Venus, at −3.9, approach to within half a degree of each other on 11 February. However, the two celestial bodies are less than 11° away from the Sun at the time and both planets are deep in the dawn twilight. Mercury joins in two days later but again, the planets are lost in light morning skies. Observers in the southern hemisphere will have the best opportunity to see these planets, as well as Jupiter's close approach to fourth-magnitude star Theta (θ) Capricorni on 18 February at around 06:30 UT when the two celestial bodies are a mere 0.03° apart. By the end of the month Jupiter is rising well ahead of the Sun as seen from southern latitudes.

SATURN was at conjunction late last month and reappears ahead of the Sun in the dawn sky by the end of February. It is found in the constellation of Capricornus, shining at magnitude +0.7. Its conjunction with the much brighter Venus occurs on 6 February but the two planets are only 12° away from the Sun at the time.

URANUS is a sixth-magnitude object in Aries and is visible in the evening sky. Viewed from northern temperate latitudes, the green planet does not set until around midnight but observers in the southern hemisphere have barely an hour after the end of twilight to catch a glimpse of it.

NEPTUNE is found in the constellation of Aquarius all year, shining at eighth magnitude. It is setting ever earlier in the evening as it approaches conjunction with the Sun next month and is soon lost in the glow of twilight. Look for it in the west at the beginning of the month.

2I/Borisov – Interstellar Comet

Neil Norman

When the discovery of Oumuamua was made back in October 2017, the rule books had to be re-written; this was the first object that had definitely originated from outside our Solar System. No longer were these objects hypothetical, they really existed. Sadly, Oumuamua was discovered after perihelion and rapidly raced away from the Sun meaning even the largest telescopes lost it within five months and very little was learned from it.

Fast forward to 30 August 2019, and the astronomy community became interested in an object that was placed on the Minor Planet Center's Near-Earth Object Confirmation Page (NEOCP) and designated gb00234. It had been discovered by Gennadiy Vladimirovich Borisov using his own designed and built 0.65-metre telescope that resides at his personal observatory, MARGO located at Nauchnyi in the Crimean Peninsula. He saw the object moving in a direction that was slightly different from the main asteroids, and informed the MPC database and discovered it was a new object. He then measured its near-Earth rating (calculated from various parameters) and it turned out to have a value of 100% or in other words, dangerous. He quickly alerted the World Webpage for confirmation of dangerous asteroids but noted that his object appeared diffuse visually and so must be a comet and not an asteroid.

Gennadiy Vladimirovich Borisov with the 0.65-metre telescope that he designed and built and which was used by him to discover 2I/Borisov. (Julia Zhulikova/Gennady Borisov)

Initial belief was that it was close to the Earth as it was moving at 40 km/s relative to the Sun and was at a distance from Earth of just 1.4 AU. The orbital period was believed to be a year. After twelve days and around 150 observations later, NASA's Jet Propulsion Laboratory's Scout page gave the object an orbital eccentricity value of between 2.5 and 4.5, which were unheard of values.

Eccentricities of Cometary Orbits

A comet with an orbital eccentricity of 0 would have a perfectly circular orbit. Those with values of up to 0.9 have stretched elliptical orbits, resulting in large orbital periods. An eccentricity value of 1 indicates a parabola, and a comet with a very large orbital period which is only loosely bound to the Sun. Comets with values >1 have hyperbolic orbits, and in these cases ejection from the Solar System may well be expected. The largest eccentricity known for a comet before 2I/Borisov was that of comet C/1980 E1 (Bowell). A value if 1.2 for this object was brought about through a close approach to Jupiter, which provided the comet with a very large gravity assist resulting in it reaching this ejection value).

This false colour image of comet 2I/Borisov was captured by the Hubble Space Telescope on 12 October 2019 when the comet was 260 million miles (418 million kilometres) from Earth. The image seen here is a composite of several exposures acquired by the Wide Field Camera 3 (WFC3) imager aboard the Hubble Space Telescope. (NASA, ESA and D. Jewitt (UCLA))

At the time of discovery, 2I/Borisov was at a low solar elongation, meaning that the Sun and comet appeared close together as seen from the Earth. This would affect the data gathered by other observers, and certainly no big telescopes would venture there for the fear of damaging their optics. However, Gennadiy employs the technique of observing as close to the Sun as possible in order to discover his comets, which at the time of writing total nine.

More observations eventually came in, and it was then determined that the comet was indeed following a very hyperbolic trajectory and was not close to the Earth at all, but was in fact beyond the orbit of Mars. The eccentricity was given as 3.3 and a perihelion distance of just over 2 AU from the Sun was attained by the comet on 8 December 2019. The designation for the comet was then changed to 2I Borisov (2I denoting the second interstellar object, and Borisov following the tradition of using the name of the discoverer).

As can be imagined, frantic observations of this comet were carried out by those observers whose instruments allowed them to reach down to its magnitude of 18. Within a matter of days the first papers had been written, including one penned by David Jewitt and Jane Luu which was the first to

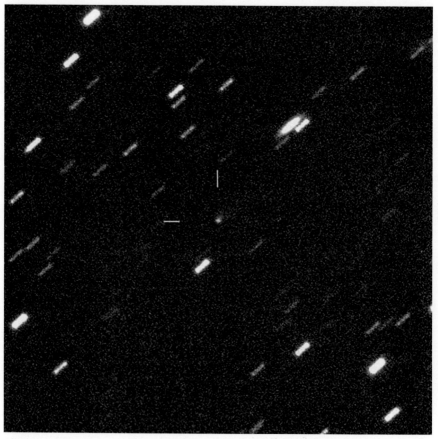

Nick James captured this image of 2I / Borisov from Chelmsford, Essex, England on 30 September 2019. (Nick James)

determine that the comet was losing dust and water at a rate of 2 kg/s and 6 kg/s respectively. Jewitt and Luu also concluded that the comet became active in June 2019, when located at a distance of between 4 and 5 AU from the Sun, albeit unobservable due to it being behind the Sun as viewed from Earth.

The size of the nucleus was first put forward by Karen Meech at the Institute of Astronomy of the University of Hawaii, who estimated a size of between 2 and 16 kilometres, whilst another paper by Piotr Guzik and others, suggested a nucleus of just 0.5 kilometre with the assumption that the albedo of the comet nucleus was 0.04 (a value typical of most comets). An improved estimate was published on 26 September by Alan Fitzsimmons and Karen Meech, after measurements of other molecules had been gathered, indicating a diameter of between 1.4 and 6.6 kilometres.

When dealing with an interstellar object, it is obviously vital to obtain a spectrum of the comet to see what it contains and to compare it with comets in our own Solar System. A low-resolution spectrum was obtained with the Gran Telescopio Canarias on 13 September, which showed that the chemical composition was not far removed from those of comets from the Oort Cloud. Further measurements obtained with the Nordic Optical Telescope found that Borisov had colour indexes (numerical expressions that determine the colour of an object) resembling those of long-period comets from the Solar System.

The William Herschel Telescope reported the detection of cyanide (CN) emissions at 388nm (typical of Solar System comets). The Bok Telescope and MMT Telescope in Arizona, USA, have reported the detection of diatomic carbon (C_2), again a very well known cometary element. The ratio of C_2 to CN is less than 0.3 which resembles carbon depleted comets in our Solar System found in the Jupiter-family comets.

This is an interesting finding as it may indicate that 2I/Borisov, when in its original home system, was once like our own Jupiter-family comets and that it may have made several perihelion passages around its parent star before encountering a large planet and ending up being ejected from the system through gravitational perturbations.

2I/Borisov entered the Solar System from the general direction of Cassiopeia, near the border with Perseus, which indicates an origin near the galactic plane. The trajectory as seen from Earth saw it as a northern hemisphere object from September, before it crossed the celestial equator on 13 November. It will leave the Solar System in the general direction of Telescopium.

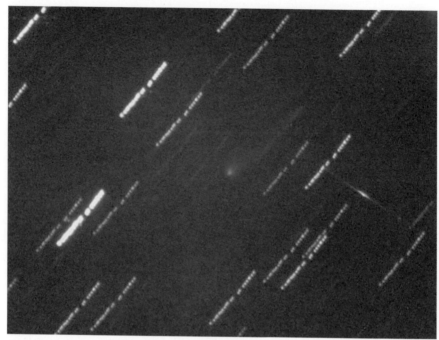

A meteor is seen passing through the field of view on this image of Comet 2I/Borisov taken by Norbert Mrozek on 4 October 2019. (Norbert Mrozek)

Although the origin of Comet 2I/Borisov remains a mystery and we may never truly know where it originated from, a candidate has been suggested. This is the binary star system Kruger 60 which lies a little over 13 light years from Earth in the constellation of Cepheus. The comet would have passed that system around a million years ago at a distance of 5.7 light years and at a low relative velocity. However, this is far from conclusive and is remains pure speculation until more detailed data is gathered.

To date, 2I/Borisov has shown many of the signs of a regular Solar System comet: no thrills, no spills. So, is there nothing more that a huge anticlimax to this object? In the writer's opinion, not at all! I believe this strengthens the view that other star systems have the same chemical makeup as we do. Indeed, armed with our new-found knowledge, we can perhaps step up the hunt for extraterrestrial life, which I believe is out there. We just have to look that little bit harder ...

March

MERCURY and Jupiter have a close pairing on 5 March when Mercury appears just 0.3° north of the much brighter gas giant. The largest greatest elongation west this year (27.3°) occurs the following day. The smallest planet in the solar system has a close encounter with Neptune on 29 March when the two bodies are 1.3° apart but Neptune requires dark skies and a telescope to see it, so this event may not be visible. Mercury is found in the morning sky and is best seen from southern and tropical regions; this morning apparition is the worst of the year for northern temperate latitudes. It plummets back toward the eastern horizon from early March, brightening slightly as it goes from crescent to gibbous phases and recedes from Earth.

VENUS has long since vanished from view for observers in northern temperate latitudes and soon gets too close to the horizon for anyone else to see this month. The morning star calls on Neptune on 14 March with the two planets appearing only 0.4° apart but this is only three days after Neptune's conjunction with the Sun so the event is unobservable. Venus itself undergoes superior conjunction on 26 March, to reappear in the west after sunset next month.

EARTH reaches an equinox on 20 March, when the Sun, travelling on the ecliptic, crosses the celestial equator. In this case, the Sun is moving from south to north, ushering in spring in the northern hemisphere and introducing autumn to the southern hemisphere. Astronomers refer to this equinox as the Vernal Equinox. Four days later, the waxing gibbous Moon is found 2.6° north of M44, the open star cluster located in Cancer.

MARS is located in the distinctive V-shaped constellation of Taurus, the Bull. It passes less than 3° south of the Pleiades open star cluster (M45) on the third day of the month in an event visible during the evening hours. The waxing crescent

Moon gets even closer on 19 March when it glides 1.9° south of the red planet. Mars is best viewed from northern latitudes where it remains aloft until after midnight, whereas observers in the southern hemisphere lose the planet about three hours after sunset. Mars continues to dim on it way to solar conjunction in October, beginning the month at magnitude +0.9 and losing nearly half a magnitude by April.

JUPITER is a morning sky object in Capricornus, rising just after midnight for observers in the southern hemisphere but only just ahead of the Sun for planet chasers in northern temperate latitudes. Mercury makes another pass by Jupiter on 5 March when the two planets are only 0.3° apart approximately 27° away from the Sun. At −2.0, Jupiter is nearly two magnitudes brighter than Mercury.

SATURN is seen in the morning sky and is best viewed from the southern hemisphere where it rises well ahead of the lightening dawn sky. At magnitude +0.8, only Jupiter and the Moon are brighter celestial objects in Capricornus.

URANUS is approaching solar conjunction next month and becoming increasingly difficult to see in the west after sunset. The skies are still dark when Uranus, located in the constellation of Aries, sets for northern hemisphere observers but the planet is lost in the evening twilight by the end of the month for planet watchers in southern latitudes. The waxing crescent Moon passes within 3° of Uranus on 17 March.

NEPTUNE is at conjunction with the Sun on 11 March and is not visible this month. A planetary conjunction with Mercury on 29 March will probably be unobservable with the two bodies being only 18° away from the Sun.

Alfred Fowler

Jane Callaghan

Few people have heard of astrophysicist Alfred Fowler, yet his achievements are astonishing.

His parents were illiterate textile mill-workers who lived in Wilsden, near Bradford in the West Riding of Yorkshire. Alfred, the sixth of their seven surviving children, was born on 22 March 1868. His private life was destined to carry more than its share of tragedy, his nearest-in-age brother dying when Fowler was just four years old. Their father subsequently suffered from severe depression, taking his own life on the day after Alfred's eighteenth birthday. In 1892 Alfred married Isabella Orr, the couple going on to have two children, Hilda and Norman, and just one grandchild, Barbara, whose own daughter Patricia (Fowler's only great-grandchild) was tragically killed in a road accident aged 21.

However, Alfred was fortunate enough to have been born on the cusp of educational reform. His phenomenal early talent for mathematics led the family to move to Keighley in 1896, improving his chances of earning a scholarship to the new Keighley Trade and Grammar School, which he duly did.

In 1882, aged only 14, Fowler gained an 'exhibition' to study mechanics at the recently-established Normal School of Science (later renamed the Royal College of Science and now known as the Imperial College of Science, Technology and Medicine). In order to further fund his place there, he became a part-time 'computer'

Alfred Fowler c1888 around the time he became Demonstrator in Astronomical Physics under Sir Norman Lockyer at the Normal School of Science. (Theresa Higgins)

to the scientist and astronomer Norman Lockyer, Professor of Astrophysics (and Head of the Solar Physics Observatory attached to the college), and then continued as full time assistant after getting his first class pass in mechanics.

Their relationship would have sorely tested a less modest man than Fowler. He later wrote "Lockyer rarely allowed me to forget that I was his assistant … [and] was not always easy to please but my own relations with him were seldom otherwise than agreeable, and I have always felt grateful that I came under his influence."

In 1901, Lockyer left the college, taking with him all the astronomical equipment to the Solar Physics Observatory of which he was still Director. Fowler was left in charge of what astrophysics remained in the college. Herbert Dingle, a colleague, said that the college astrophysics department in 1901 comprised "a table in a frequently used lecture theatre, in the corner of which a dark room had somehow to be improvised. For apparatus he [Fowler] had one small spectrograph of a type we should now consider suitable for a promising child's stocking at Christmas; and for encouragement the collected indifference of most of those who might have been concerned."

Alfred Fowler in 1931 in his office at Imperial College. Behind him, on the walls, are framed photographs of spectrum bands .The microscope in front of him is evidence of the eye-straining work involved in analysing the photographic film. (Theresa Higgins)

Pragmatically, Fowler concentrated on the laboratory work and gradually developed the spectroscopy department into a world leader at a time when the Lick Observatory in California was significantly better funded and equipped. He was an acknowledged expert in solar and stellar spectroscopy, being one of the first to determine that the temperature of sunspots was cooler than the surrounding photosphere.

Alfred Fowler was elected a Fellow of the Royal Society in 1910, and awarded their Royal Medal in 1918 in recognition of his research into physical astronomy

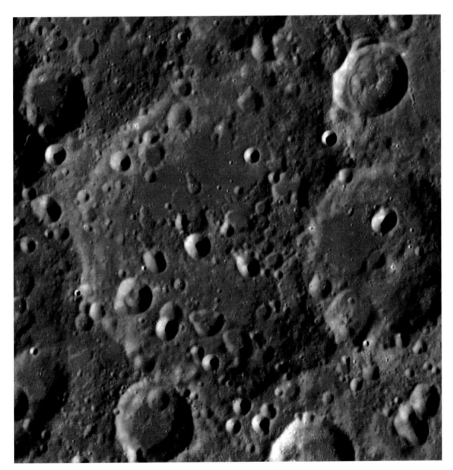

The heavily worn lunar crater Fowler, pictured here with the smaller but more distinct impact crater Von Zeipel overlying its eastern rim. (NASA/LROC)

and spectroscopy. This was followed, despite his dislike of being centre-stage, by his becoming the first General Secretary of the International Astronomical Union, and serving as President of the Royal Astronomical Society from 1919 to 1921. He was awarded prize after prize, culminating in the CBE in 1935.

In 1934 Fowler retired in fragile health and died just six years later in Ealing, Middlesex on 24 June 1940. Today this once world-renowned expert in stellar spectroscopy is largely forgotten. His obituary said: "Few men have seen so large a proportion of their work become a common possession of which not many remember the source. He was content that it should be so. His satisfaction came from the consciousness of work well and truly done, that would outlive his own name."

Variously described as "amiable", "shrewd" and "inspiring", Alfred Fowler had, according to his colleague Henry Plummer, a "sturdy character which carried him unruffled through the difficulties and disappointments of life". Indeed, his many achievements have been recognised by the 146 kilometre diameter lunar crater Fowler, located in the northern hemisphere of the far side of the Moon, being named after him, an honour he shares with the English physicist and astronomer Ralph Howard Fowler after whom the crater is also named.

April

New Moon: 12 April
Full Moon: 27 April

MERCURY is still high enough in the east to be seen from tropical and southern latitudes but it is plummeting toward the Sun and is gone by mid-month. Favoured observers might catch a glimpse of the very old crescent Moon 3.0° south of the magnitude −1.2 planet on 11 April but the two objects are a scant 10° from the Sun at the time. Mercury is at superior conjunction just eight days later and then reappears in the evening sky at sunset. The northern hemisphere will then get the best views of the closest planet to Sun although observers will be hard-pressed to witness Mercury passing 0.7° north of sixth-magnitude Uranus on 24 April.

VENUS appears low in the west at evening twilight, a small waning gibbous disk as seen through a telescope, shining at a steady magnitude −3.9. This is the beginning of an excellent apparition for southern hemisphere astronomers and quite a good one for observers in equatorial regions; however, it is thoroughly disappointing for sky watchers in the north. The New Moon is 2.9° south of Venus on 12 April but both celestial bodies are only 3° from the Sun at the time. Venus and Uranus pass within 0.2° of each other on 23 April but with the event occurring only 7° from the Sun and in twilight skies, it is probably unobservable. The slightly more distant Mercury–Venus conjunction two days later is similarly difficult to see due to the proximity of the Sun.

EARTH enjoys another major meteor shower this year when the Lyrids put on a show in the latter half of the month. The article *Meteor Showers in 2021* contains more information. The Moon occults Mars on 17 April and passes less than 3° north of Praesepe (M44) three days later. The Full Moon of 27 April occurs only 12 hours before perigee, raising the possibility of extreme tides on Earth.

MARS continues to inhabit the evening sky, dimming from magnitude +1.3 to +1.6 over the course of the month. It spends most of April in the constellation of Taurus, entering Gemini on 24 April. Mars reaches this year's maximum declination north of +24.9° the day before that. Observers in India and southeast Asia will have an opportunity to watch the waxing crescent Moon occult the red planet on 17 April, beginning around 13:00 UT.

JUPITER is best seen from the southern hemisphere where it rises around midnight. Astronomers in northern latitudes continue to find Jupiter quite low in the brightening eastern sky, rising only an hour or so ahead of the Sun. Jupiter occults the sixth-magnitude star 44 Capricorni on 2 April. Beginning around 08:45 UT, the event lasts approximately 75 minutes. On 25 April, Jupiter leaves Capricornus behind and enters the neighbouring constellation of Aquarius where the gas giant remains until August.

SATURN continues slowly moving across the constellation of Capricornus this month. Shining at magnitude +0.8, Saturn remains mired in morning twilight for observers in northern temperate latitudes. However, the view from the southern hemisphere is far better, with the ringed planet rising before midnight.

URANUS has another close encounter with our satellite this month, with the very young crescent Moon gliding less than 3° south of the planet on 13 April. However, this event will be difficult to see as both celestial bodies will be only 15° away from the Sun and deep in the bright evening twilight. The faint planet arrives at conjunction with the Sun on the last day of the month and is unobservable for much of this month and next.

NEPTUNE was at conjunction with the Sun last month and is just reappearing in the morning sky. It remains lost in the brightening dawn sky of the northern hemisphere in April but is soon visible telescopically to observers in southern latitudes where the nights are lengthening.

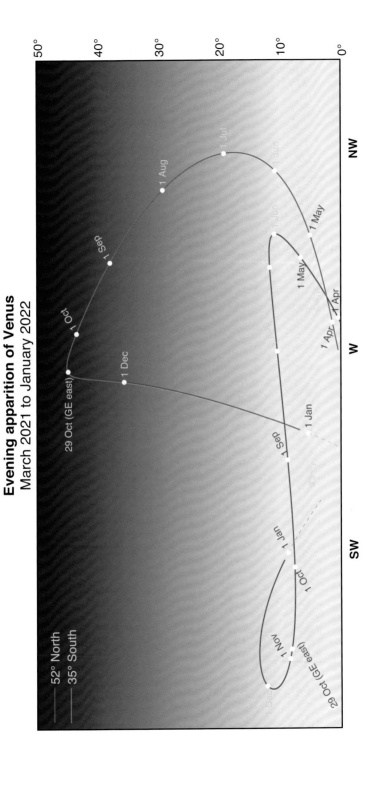

Evening apparition of Venus
March 2021 to January 2022

52° North
35° South

1 Jul
1 Aug
1 Sep
1 Oct
29 Oct (GE east)
1 Dec
1 Jun
1 May
1 Apr
1 Jun
1 May
1 Apr
1 Aug
1 Jan
1 Sep
1 Jan
1 Oct
1 Nov
29 Oct (GE east)
1 Dec

NW
W
SW

0°
10°
20°
30°
40°
50°

Salyut 1: The First Space Station

Neil Haggath

This April sees the 50th anniversary of the world's first space station.

After the Soviet Union had lost the race to the Moon, its space efforts turned to the development of space stations, to study the effects on humans of long-term spaceflight. Its first space station, Salyut 1 – meaning "Salute" – was launched on 19 April 1971. It had been intended to launch it a week earlier, on the tenth anniversary of Yuri Gagarin's flight, but technical problems caused the delay. The intention was for successive three-man crews to live aboard the station, ferried to and from it by Soyuz spacecraft.

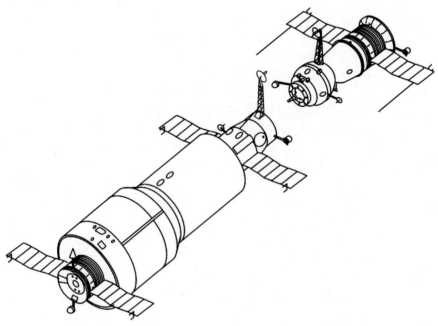

A Soyuz spacecraft (right) preparing to dock with Salyut 1. Note the station's Soyuz-type solar panels, and at its rear (left), its Soyuz-based propulsion module. (Wikimedia Commons/NASA/David S. F. Portree)

Salyut 1 was much smaller than the USA's 90-ton Skylab, which followed two years later. It was 12 metres long, with a maximum diameter of 4 metres, and a mass of 16.3 tons. With a Soyuz docked to it, its total length was 20 metres, and its mass 22.7 tons. The Soyuz was not merely a ferry, but was an integral part of the station, as its orbital module was used as the sleeping accommodation.

The station had eight working positions, some for control of the station itself, and some for operating various experiments. It carried several astronomical instruments, including a telescope and spectroscope, and equipment for medical and biological experiments. One of its objectives was to perfect the design of such instruments for use on future stations. It remained in orbit for nearly six months and 2,800 orbits.

At 222 × 200 km, Salyut 1's orbit was so low that it had to be frequently raised, otherwise it would have re-entered the atmosphere in about a week. For this purpose, it had its own rocket engine, developed from that of a Soyuz.

The first intended crew, cosmonauts Vladimir Shatalov, Alexei Yeliseyev and Nikolai Rukavishnikov, were launched aboard Soyuz 10 three days after the station was launched. They successfully docked with Salyut, but were unable to enter it; after five and a half hours docked to it, they separated and returned to Earth. The official explanation was that a technical problem prevented them opening the hatch, although it was also speculated that Rukavishnikov, the crew's only first-timer, had become ill. It was reported, however, that they docked and undocked more than once, to practice the technique.

Six weeks later, on 6 June, Giorgi Dobrovolsky, Vladislav Volkov and Viktor Patsayev, the crew of Soyuz 11, successfully docked with and boarded Salyut 1. They set what was then an endurance record of 23 days 18 hours in space, all but about one day aboard the station. But their mission ended in tragedy; during their return to Earth on 30 June, the crew – who flew without spacesuits – were killed by a catastrophic leak and loss of cabin pressure. This tragic event is described in detail in the author's *Anniversaries in 2021* article elsewhere in this Yearbook.

Salyut 1 was kept habitable, and continued to fire its engine weekly to raise its orbit, for a further three months, with the intention of another crew visiting it. But presumably the investigation into the accident, and the subsequent modification of the Soyuz spacecraft, prevented that happening. By early October, Salyut was running out of fuel, and on 11 October, it was deliberately de-orbited and destroyed on re-entry. It was not allowed to re-enter due to natural orbital decay, in case pieces of it survived intact and fell onto populated areas.

After two failed attempts in 1973 – one of which failed to reach orbit, and was not given a Salyut designation – five more successful Salyut stations were flown between 1974 and 1986. Two of these were in fact military reconnaissance stations – and one even tested a space weapon – but were given Salyut names to conceal their true purpose. The last two, Salyuts 6 and 7, were far larger and more sophisticated, and lasted far longer, than their predecessors. Salyut 7 remained in orbit for nearly nine years, and was manned by ten different crews. Each of Salyut 6 and 7 was manned by successive two-man crews, who stayed aboard for durations of up to seven months, and hosted further three-man crews who made short visits. The latter included several "guest" cosmonauts from other communist bloc nations.

The Salyut programme ended with Salyut 7 in 1986, but was followed by the long-lived Mir station.

May

New Moon: 11 May
Full Moon: 26 May

MERCURY presents its best evening apparition of the year for planet watchers in the northern hemisphere. As with all evening appearances by this tiny planet, it is brightest at the beginning of the apparition (superior conjunction) and faintest at the end (inferior conjunction). This month, Mercury starts off at a bright magnitude −1.2 but ends as a third-magnitude object. Mercury calls upon the Pleiades (M45) on 4 May, gliding 2.1° south of the open cluster in the constellation of Taurus. The young crescent Moon passes a similar distance from Mercury on 13 May. Greatest elongation east, at 22.0°, occurs four days later. After this, Mercury begins to lose altitude but remains visible, albeit dimmer, until early next month. Mercury and Venus are found together in the evening twilight on 29 May when the tiny planet appears 0.4° south of the evening star. The following day, Mercury embarks upon retrograde motion.

Evening Apparition of Mercury
19 April to 11 June

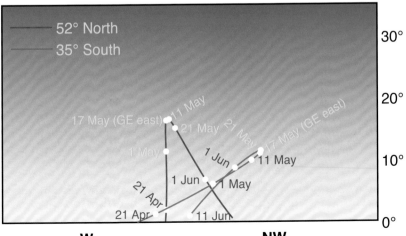

VENUS continues to gain altitude above the western horizon, the evening star being best viewed from tropical and southern latitudes. In a telescope, it still appears as a small waning gibbous disk shining at magnitude −3.9. Islanders off the west coast of South America may see the evening star occulted by the waxing crescent Moon on 12 May in an event starting around 22:00 UT. Mercury and Venus come together on 29 May when the two planets appear less than half a degree apart.

EARTH runs into the debris left behind by Comet 1P/Halley early this month, which becomes the Eta (η) Aquariid meteor shower. For more information, consult *Meteor Showers in 2021* elsewhere in this volume. The Moon is quite busy, reaching its most distant apogee of the year on 11 May, occulting Venus the following day, passing by Praesepe (M44) on 18 May and participating in a total lunar eclipse on 26 May. The section *Eclipses in 2021* contains all of the details for observing this event. The Full Moon of 26 May is also a 'Super Moon' as it exhibits the largest angular diameter of the year.

MARS briefly occults the sixth-magnitude star HD 47020 on 7 May in an event beginning at approximately 23:15 UT and lasting just minutes. The red planet and the Moon have a close encounter nine days later although an occultation does not take place when the waxing crescent Moon glides 1.5° north of the red planet. Mars is located in Gemini in June, dimming slightly from magnitude +1.6 to +1.7 by the end of the month. The red planet is an evening sky object, perhaps most easily observed from the northern hemisphere where it sets around midnight. Mars aficionados in southern latitudes need to look for the planet as soon as it gets dark as they have three hours or less in which to observe the object.

JUPITER reaches west quadrature on 21 May. With the shadows of the planet and its moons cast slightly off to one side, this is an interesting time to watch the Jovian system through a telescope. Now located in Aquarius, Jupiter is best seen from the southern hemisphere where it rises late in the evening. It is still strictly a morning-sky object for northern viewers, however, although it is slowly distancing itself from the Sun.

SATURN rises in mid-evening for observers in the southern hemisphere; this is the best place to be to see the ringed planet. The constellation of Capricornus is low in the spring skies of northern temperate latitudes so Saturn is a difficult target for planet watchers in the north. Saturn arrives at west quadrature on the third day of the month and reaches its maximum declination north ($-17.4°$) for the year on 18 May. Two days later, the rings close to their minimum opening angle, $16.7°$, of 2021. Saturn reverses course on 23 May, going from direct to retrograde motion.

URANUS was a solar conjunction late last month and is not visible in the morning sky for much of this month. It is best viewed from southern latitudes where it emerges from the dawn sky near the end of May. The very old crescent Moon passes near the green ice giant on 10 May but both bodies are too close to the Sun for this event to be seen.

NEPTUNE remains a difficult object in northern temperate latitudes where the sky never achieves true darkness. Neptune is only magnitude +7.9; thus, a small telescope (and no light pollution) is necessary to glimpse it. Observers in the southern hemisphere have the best opportunities to see the most distant planet in the solar system this month. It rises around midnight by the end of May and is well aloft by sunrise. Look for it in Aquarius.

Thomas Henry Espinell Compton Espin
The Double Star Curate of Tow Law

John McCue

Brilliant or bossy? Sheer genius or chauvinist? The Rev. Thomas Henry Espinell Compton Espin, M.A. would certainly be a controversial figure in the twenty-first century.

Born in Birmingham on 28 May 1858, his interests and commitments were many and varied and included those of: "... priest, musician and composer, inventor, radiographer, astronomer, microscopist, botanist and geologist …"[1] although not necessarily in that order! Thomas was educated at Haileybury public school where the Rev. John Hall, one of his teachers, instilled in him an astronomical interest, the flames of which were fanned by the appearance of Coggia's Comet in 1874. Also known as the Great Comet of 1874, this non-periodic comet was discovered on 17 April 1874 by French astronomer Jérôme Eugène Coggia at the Marseille Observatory and attained a maximum magnitude of between 0 and 1 in July of that year.

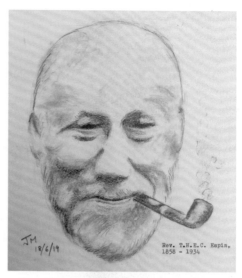

Espin with his ever-present pipe. He smoked until the bowl was caked and blocked inside, and then discarded it for a new one! (John McCue)

Thomas' early observing was carried out with opera glasses, after which he progressed to one-inch and then three-inch refractors. He sent his observations to the *English Mechanic*, a renowned science magazine of the time. When Espin was a mere seventeen years of age, he was asked by none other than Thomas William Webb to help him produce his iconic *Celestial Objects for Common Telescopes* which was destined to become a favourite with amateur astronomers – the last edition in 1962 is still available and well-loved.

Espin attended Oxford and while there he found his first double star with his three-inch refractor, bringing him to the attention of Professor Charles Pritchard. The academic allowed Espin to use the 13-inch university refractor, on condition that he helped the professor with his telescopic work and tutored students in astronomy.[2] In 1881, Espin graduated from Oxford and shortly afterwards appointed as a curate in the Wirral. While there, an advertisement in the aforementioned *English Mechanic* magazine led to him helping form the Liverpool Astronomical Society.

1. Anthony Driver, 'Vicar of Tow Law, 1982–1991', foreword to *The Stargazer of Tow Law*.
2. K.L.Johnson, *Oxford Dictionary of National Biography*.

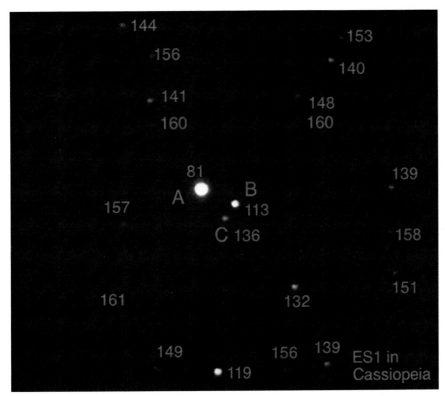

A photograph of ES1, taken by Christopher Walker during a visit to his observatory on the Mull of Galloway, Wigtownshire, Scotland with an Altair GPCAM2 224C astro-camera on a 14-inch Newtonian reflector, f/4.5, exposure 1 second, auto ISO. The numbers show the apparent magnitudes of the stars (decimal points omitted). (Christopher Walker)

During his astronomical career Espin discovered 2,575 new double stars, which bear his code ES in the *Washington Double Star Catalogue*, the repository of all double star measurements across the globe. ES 1 is in the constellation of Cassiopeia, the two stars forming it being of magnitudes 8.1 and 13.6 and separated by 22 arc seconds. The optical double ES 2575 nestles in Cygnus and comprises two stars of identical magnitude 10.6 and separated by only 4 arc seconds. In addition, Espin found 3,800 unusual red stars; noted over 30 new variable stars; used a spectroscope of his own design to observe the spectra of every star brighter than ninth-magnitude (numbering over 100,000 in total

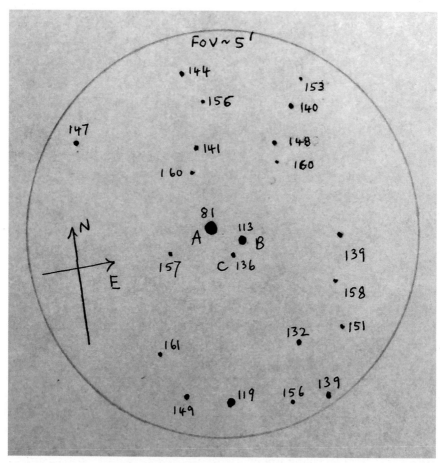

This sketch of ES1 is based on the image shown above and aims to emphasise the stars. The original image was processed and analysed to measure the position angles and separations (not shown here). The apparent magnitudes of the stars are shown and correspond to those in the photograph. (John McCue)

in Argelander's Charts); and successfully submitted over 1,000 papers to the *Monthly Notices of the Royal Astronomical Society.*[3]

However, his greatest astronomical achievement was probably the discovery of Nova Lacertae on 30 December 1910, (when the vast majority of the

---

3. The Tow Law History Group, *The Stargazer of Tow Law*, foreword by Patrick Moore, 1992.

population would no doubt be getting ready for the New Year celebrations!). This star is now designated DI Lacertae and was shining at magnitude 4.6 when Espin first spotted it, after which it faded by three magnitudes over the following 37 days. Its spectrum displayed bright emission lines from hot gases, which Espin may well have seen with the help of his homemade spectroscope. Currently at magnitude 14, ultra-violet spectral analysis suggests the existence of an accretion disk around a white dwarf.[4]

By the time of that momentous discovery, Espin was the well-established curate in the town of Tow Law in Durham, having been appointed in 1888. He had transported his 17¼-inch reflector, built by the famous East Anglian telescope maker George Calver, to this new location from Wolsingham, where Espin had been second curate at his father's parish for the previous three years. He followed this up with a 24-inch reflector (also built by Calver) in 1914. In 1961 David Sinden, a telescope maker at Grubb Parsons, Newcastle, rescued this fine instrument from thirty years of decay and hen droppings and relocated it to an observatory provided by Newcastle University. Unfortunately, the telescope has once again fallen into disuse.

Espin had a wide and varied outlook on life, and he lived it to the full. A remarkable adventure is recounted[5] of his holiday to Italy with his assistant William Milburn. They visited and studied the active volcano Vesuvius. Milburn wrote: "We ascended to the top by rack and pinion railway, and after a short walk came to the edge of the crater ... the volcano was throwing out volumes of steam and showers of stones from its centre and pouring lava out to the north ... We struck off directly for the eruptive cone, and heard the ejected stones continuously falling around us. An alarming feature was that apparently solid rock broke under your feet, and your leg slipped well in while steam poured out of the hole. We walked over the still hot lava and ascended the cone to within a few yards of the top. It was too risky to venture further ... Forty days afterwards the cone blew up and turned the whole basin into a lake of fire."[6] At one point during this holiday, both men were overcome by fumes and

4. Elizabeth Moyer et al, (December 2007), 'Hubble Space Telescope Observations of the Old Nova DI Lacertae', *The Astronomical Journal*, 125, 288.
5. The Tow Law History Group, *The Stargazer of Tow Law*, foreword by Patrick Moore, 1992.
6. 4 W. Milburn, *Northern Weekly Gazette*, 10 May 1924.

This Apollo 16 image shows the 75 kilometre diameter lunar crater Espin. Located just beyond the north eastern limb on the far side of the Moon, the crater was named in honour of Thomas Henry Espinell Compton Espin. (NASA/James Stuby)

had to be rescued. It is hard to imagine any tourists being allowed to risk their lives in such a fashion if these circumstances had occurred in modern times.

Returning to the opening question – was Espin brilliant or bossy? Sheer genius or chauvinist? Maybe all four! During his time at Tow Law, his fierce independent spirit refused to countenance a parish council – all church business was done by him alone. He also refused to allow women into the church choir. Nonetheless, he "abounded in human sympathy, as was indicated by all he did for his parishioners, both sick and whole".[7] Stories of his medical help (he built his own X-ray machine) for unwell Tow Law people abound.

Thomas Henry Espinell Compton Espin died at Tow Law on 2 December 1934, aged 76.

7. The Tow Law History Group, *The Stargazer of Tow Law*, foreword by Patrick Moore, 1992.

June

New Moon: 10 June
Full Moon: 24 June

MERCURY bows out of the west early in the month and undergoes inferior conjunction on 11 June, reappearing in the east before sunrise by mid-month for southern hemisphere observers and a little later for those in the north. It passes 1.8° north of Aldebaran (α Tauri, the 'eye' of the Bull) on 22 June and returns to direct motion the following day. This is a fair morning apparition which favours equatorial latitudes. The brightness of Mercury varies from sixth magnitude at inferior conjunction to first magnitude by the end of the month, but it will continue to brighten throughout July.

VENUS continues its ascent in the western sky as viewed from equatorial and southern latitudes. Observers in northern temperate latitudes, however, need a low horizon as the magnitude −3.9 evening star barely clears 10° in altitude before starting to descend again. The very young crescent Moon appears 1.5° north of Venus on 12 June, an attractive pairing in the evening twilight.

EARTH experiences an annular solar eclipse on 10 June, the details of which may be found in *Eclipses in 2021*. The dark skies during the nights either side of the eclipse can only enhance observations of the Tau (τ) Herculid meteor shower; more information is in the article *Meteor Showers in 2021*. Solstice takes place on 21 June when the Sun reaches its most northerly declination of the year. This day marks the beginning of summer in the northern hemisphere and the onset of winter in the south. A week before that, the waxing crescent Moon passes just under 3° north of the open cluster Praesepe (M44) in Cancer.

MARS continues its presence in the evening sky although it is setting earlier every night. The red planet disappears in the early evening; this is during the evening twilight for northern observers but the skies will be properly dark from southern latitudes. Mars moves from Gemini to Cancer on 8 June and passes just 0.3° south of M44 on 23 June. The thin waxing crescent Moon passes just under 3° north of Mars on 13 June.

JUPITER reaches its maximum declination north of −11.7° for the year on 16 June. Best seen from the southern hemisphere where it rises mid-evening, the gas giant is now appearing in the east around midnight for northern observers. At magnitude −2.5, Jupiter is the brightest star-like object in the otherwise faint constellation of Aquarius. However, its tenure in the 'Water Carrier' will soon be coming to an end; Jupiter goes into retrograde motion on 21 June and heads back in the direction of Capricornus.

SATURN rises in the early evening hours for viewers in southern latitudes. However, the ringed planet is only just appearing in the east before midnight for planet watchers in the northern hemisphere. Saturn is brightening as it approaches opposition in August. Also, the rings are opening slightly, adding to the planet's overall apparent magnitude of +0.6 in the constellation of Capricornus.

URANUS is still embedded in morning twilight at the beginning of June for observers in northern temperate latitudes but it is readily accessible in the dark skies of the southern hemisphere. The waning crescent Moon moves past the dim planet on 7 June, getting a little closer every month until November. Look for Uranus in the faint constellation of Aries.

NEPTUNE remains primarily a morning sky object but is finally rising before midnight for planet chasers in the southern hemisphere. The long winter nights allow for plenty of observation time of this eighth-magnitude telescopic object in the zodiacal constellation of Aquarius. However, Neptune continues to be a difficult object for observers in northern temperate latitudes due to the lack of a truly dark sky. The blue ice giant reaches west quadrature on 13 June and its annual maximum declination north (−3.9° in 2021) nine days later. On 26 June, Neptune arrives at a stationary point and reverses course across the background stars.

The Star That Was Older Than the Universe
The Mystery of HD 140283

David Harper

A Crisis in Astronomy?

In the summer of 2019, tabloid newspapers ran stories about a crisis in astronomy. Scientists attending a cosmology conference in California were shocked to discover that a nearby star was older than the universe itself! The culprit was a seventh-magnitude star in Pisces, designated HD 140283 in the Henry Draper Catalogue and nicknamed the Methuselah Star because of its extreme age.

A Very Peculiar Star

HD 140283 had been recognised as an unusual star as early as 1912, when Walter Adams measured its radial velocity using the 60-inch reflector at Mount Wilson and found that it was approaching the Sun at 170 kilometres per second. The Mount Wilson Spectrographic Parallax Catalogue of 1935 listed it as a white dwarf with a classification of A5sp. The suffix letters "sp" denote very sharp (narrow) lines in the spectrum and peculiar metal abundances, respectively. The star was observed again by Joseph Chamberlain and Lawrence Aller in 1950, using a spectrograph on the 100-inch telescope at Mount Wilson. They discovered that it appeared to have much less iron and other heavy elements than the Sun. This observation provided important supporting evidence a few years later when Margaret Burbidge and colleagues published their pioneering paper on the nucleosynthesis of elements in stars.

The Oldest Star

Modern observations of HD 140283 show that its iron-to-hydrogen abundance is 1/250th of the value measured for the Sun. Astronomers classify it as a Population II sub-giant star. These are the oldest stars that are still observable. They are mainly found in globular clusters and the galactic halo, and they have characteristically low abundances of elements heaver than hydrogen and helium. The Sun, by contrast, is classified as Population I, a star which is relatively young and rich in such elements. The first stars to form after the Big

A Digitized Sky Survey image of HD 140283 (centre) and its surrounding star field. The field of view is 25′ x 20′ and the faintest stars are 20th magnitude. (Digitized Sky Survey (DSS), STScI / AURA, Palomar / Caltech, and UKSTU / AAO)

Bang are known as Population III. They contained only hydrogen and helium, because these were the only elements that existed immediately after the Big Bang, and they were very massive and very luminous, living for only a few million years before ending in supernova explosions which seeded the early universe with its first heavy elements. It was from the debris of these first stars that HD 140283 gained its meagre amounts of iron.

HD 140283 lies just 200 light years from the Sun. In addition to its high radial velocity, it has a very large proper motion. This has led astronomers to conclude that it is a temporary visitor from the galactic halo which is just passing through the solar neighbourhood.

Measuring the Age of Methuselah

In 2013, a team led by Howard Bond at the Space Telescope Science Institute published a paper in the journal *Astrophysical Journal Letters* which reported their work to measure the age of HD 140283. They planned to do this by running computer simulations of its evolution from its birth to the present, and then comparing the simulation to the observed properties of the star. HD 140283 is a good candidate for this method, because it is at a stage of its evolution between being a main sequence star like the Sun and becoming a red giant.

Their first challenge was to determine its distance more accurately, in order to establish its intrinsic luminosity. They were unsatisfied with the parallax that had been determined by ESA's Hipparcos mission, which had a 4% margin of error, so they devised an ingenious way to use the Hubble Space Telescope's fine-guidance sensors to achieve an almost five-fold improvement, reducing the error to less than 1% of the measured value.

The computer simulations used a mathematical model for stellar evolution which was developed at the University of Victoria (Canada) by Don VandenBerg and colleagues. Among the parameters that must be entered into the mathematical equations at the start of the calculation is the amount of oxygen present in the star at its birth. This is usually represented as the logarithm of the oxygen-to-hydrogen ratio compared to that of the Sun, and is denoted by $[O/H]_0$. The star's present value of $[O/H]$ can be measured using spectroscopy, and this can be adjusted to give the value at the star's birth. However, there is a degree of uncertainty in any astronomical observation, which means that the value of $[O/H]_0$ is not known exactly.

The result of the computer simulations was a series of Hertzsprung-Russell diagrams of luminosity versus surface temperature overlaid with isochrones. These are curves which represent a population of stars of the same age. The observed luminosity and surface temperature of HD 140283 is superimposed on the diagram, and its age can be read off directly from its location relative to the isochrone curves.

Bond and his colleagues found that the age of HD 140283 was very sensitive to the values they adopted for $[O/H]_0$. Taking the most likely value, they found an age of 14.46 billion years, which exceeds the current best estimate for the age of the universe from the Hubble Constant and from spacecraft observations of the cosmic microwave background. This is 13.77 billion years. However, Bond and colleagues estimate that their age for HD 140283 has a margin of error of

A map of fluctuations in the cosmic microwave background observed by the Wilkinson Microwave Anisotropy Probe. The WMAP mission enabled cosmologists to determine the age of the universe to better than 1% precision. (NASA)

0.80 billion years, so the star could be as young as 13.66 billion years. This is still almost as old as the universe, but not so old as to lead to the paradox of a star that is older than the universe itself.

Revisiting the Oldest Star

Another team, led by Orlagh Creevey of the Observatory of the Cote d'Azur in France, was also studying HD 140283. Its unique status as the oldest known nearby star makes it a target of interest for the European Space Agency's Gaia project, so Creevey and her colleagues undertook an intense programme of observations using ground-based instruments to measure its diameter, total energy output and surface temperature. Then they attempted to determine the age of HD 140283 using computer simulations, following a similar approach to Howard Bond and his team, but employing a different mathematical model, developed by Pierre Morel and Yveline Lebreton. They reported their results in a paper in the journal *Astronomy & Astrophysics* in 2015. They found that the calculated age of HD 140283 depended on the value adopted for the star's mass, and also on the amount by which the star's light has been reddened as it passes through the 200 light years of space which separates it from the Earth. When they assumed no reddening of the starlight, the star's age was 13.7 billion years (with a margin of error of 0.7 billion years), but if they assumed a reddening

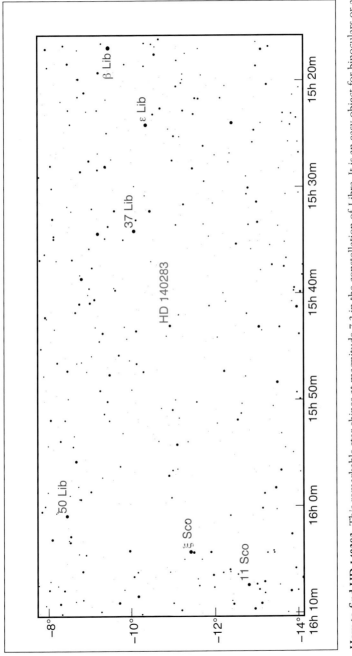

How to find HD 140283. This remarkable star shines at magnitude 7.2 in the constellation of Libra. It is an easy object for binoculars or a small telescope. This finder chart is a 12 degree by 6 degree field with HD 140283 at the centre. Stars are shown to magnitude 10, which is the limit for a good pair of binoculars in dark skies. Coordinates are referred to the epoch J2000. (David Harper)

of 0.1 magnitudes, the calculated age fell to 12.2 billion years (with a slightly smaller margin of error). They recommended that further observations should be made, to determine both the reddening and the mass more accurately.

Conclusion

There is no doubt that HD 140283 is one of the oldest stars in the universe, and astronomers will continue to study it for many years to come. It is possible that they will discover that it really is older than our current best estimate of the age of the universe. Astronomy has been here many times before. It was once believed that the Earth – and the rest of Creation – was only 6,000 years old. The pioneering geologist William Smith showed, in the late 18th century, that it must be far, far older. By the middle of the 19th century, astronomy faced a crisis when it became apparent that if (as the geological evidence implied) the Earth was many millions of years old, the Sun must be equally old, but astronomers were unable to explain how the Sun could continue shining for such a vast period of time. That puzzle was solved only in the 20th century, when nuclear fusion was revealed to be the Sun's source of energy.

Paradoxes such as stars that appear to be older than the universe may provoke hysterical headline in tabloid newspapers, but they are often the catalysts which lead to a new and deeper understanding of the universe.

Acknowledgements

I would like to thank my colleague Pete Juhas for drawing my attention to the newspaper reports about the "crisis" in astronomy which led me to research and write this article.

Further Reading

Bond, H.E., Nelan, Edmund P., VandenBerg, D.A., Schaefer, G.H, Harmer, D. (2013) 'HD 140283: A star in the Solar neighbourhood that formed shortly after the Big Bang.' *Astrophysical Journal Letters* 765:L12. (DOI: 10.1088/2041-8205/765/1/L12)

VandenBerg, D.A.. Bond, H.E., Nelan, E.P., Nissen, P.E., Schaefer, G.A., Harmer, D. (2014) 'Three ancient halo subgiants; Precise parallaxes, compositions, ages, and implications for globular clusters.' *Astrophysical Journal* 792:110. (DOI: 10.1088/0004-637X/792/2/110)

Creevey, O.L., Thevenin, F., Berio, P. and 18 co-authors. (2015) 'Benchmark stars for Gaia: Fundamental properties of the Population II star HD 140283 from interferometric, spectroscopic, and photometric data.' *Astronomy & Astrophysics* 575, A26. (DOI: 10.1051/0004-6361/201424310)

July

New Moon: 10 July
Full Moon: 24 July

MERCURY reaches a greatest elongation west of 21.6° on 4 July. Shortly afterwards, the morning sky planet begins its descent back toward the eastern horizon. Mercury brightens over three magnitudes this month, from +1.0 to −2.2 but with superior conjunction looming, the tiny planet vanishes from view before the end the month.

Morning Apparition of Mercury
11 June to 1 August

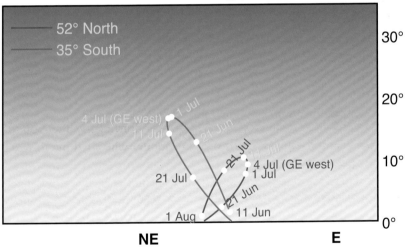

VENUS passes 0.1° north of Praesepe or the Beehive Cluster (M44) on 3 July. Ten days later, the evening star approaches the much fainter planet Mars, the two planets over 28° away from the Sun. Regulus, the brightest star in Leo, is the inferior planet's next target, with Venus appearing a degree north of the star on 21 July. The evening star appears as a waning gibbous disk in a telescope, slowly growing larger as it approaches Earth, and shining at a steady magnitude −3.9.

EARTH arrives at aphelion on 5 July, marking its most distant point from the Sun. This year, the aphelion distance is 1.107 au or 152,100,000 km. At the end of the month, the Delta (δ) Aquariids embellish the skies with faint meteors. See *Meteor Showers in 2021* for more details. The Moon also reaches First Quarter twice in July, on the first and last days of the month.

MARS leaves Cancer for the constellation Leo on 10 July. Three days later, the red planet reaches aphelion. On the same day it is also approached in the sky by the evening star, brilliant Venus, with the two planets only half a degree apart. Mars glides 0.6° north of Regulus on 29 July. This month also marks the 45th anniversary of the arrival of **Viking 1**, the first truly successful landing on Mars.

JUPITER is found in retrograde motion amongst the faint stars in the constellation of Aquarius and brightening every night as it approaches opposition next month. The gas giant rises during the early evening hours and is still above the horizon at sunrise. Jupiter celebrates its own 'Fourth of July' this year with the fifth anniversary of the arrival of the **Juno** orbital spacecraft.

SATURN is visible in the evening sky, although more difficult to spot from northern latitudes where the ecliptic remains low to the southern horizon. With opposition early next month, the ringed planet rises shortly after sunset and is aloft most of the night. Look for the magnitude +0.4 object retrograding through the constellation of Capricornus.

URANUS is found less than 2° from the waning crescent Moon on 4 July. Located in Aries, Uranus is a morning sky object, more easily seen from southern latitudes where the skies are dark. The faint planet rises before midnight for planet watchers in the northern temperate zone but this is during astronomical twilight, making observations of the sixth-magnitude object more difficult.

NEPTUNE is currently in retrograde in the constellation of Aquarius. Shining at magnitude +7.9, a small telescope is necessary to view it. It rises in the mid-evening hours for southern hemisphere observers and is best viewed in these dark skies. Observers in northern temperate latitudes are beginning to get better views as the dim planet rises ever earlier. However, planet watchers in the far north still must contend with night skies which never truly darken, making Neptune a challenging object during the summer months.

Early Precariously Balanced Refractors

Gary Yule

Over the years as an amateur astronomer, I have developed a bit of love for antique telescopes. Whether it is it their different designs and mountings, or the different materials used to make them, they have always held a fascination for me. What has always been of particular interest is the way that refractors in the early years grew in length, which begs the question as to how these great refractors were mounted and balanced in order to be good enough to use?

Telescope users are well aware of the advanced optics their instruments contain today, compared to that of what astronomers had to contend with in the early years of telescope development. The high quality optics of today yield crisp and sharp views which we now take for granted, and are often mounted on some sort of advanced robotic tracking mount that has a perfectly balanced counterweight system using clutches and gears.

Prior to the introduction of the achromatic lens in 1773 astronomers had to contend with views plagued with chromatic aberration, often referred to as "false colour". Chromatic aberration, and to the lesser extent spherical aberration, are natural defects of simple lenses. They produce views of celestial objects surrounded by unwanted colour fringing due to the inability of simple lenses to bring all colours in an image to the same focus. The result is a lower image contrast and washed out colours.

Yet these early telescopes opened up a new world which famously included the great discoveries of Galileo equipped with his small 15mm aperture lens. Later and larger versions of up to 200mm aperture were made by Christiaan Huygens, Jean-Dominique Cassini and Johannes Hevelius, who realised that extremely long focal ratios were required to correct the chromatic aberration suffered by the simple lenses available to them. They discovered that if the aperture was doubled then the focal length had to be squared (or made four times longer) in order to keep chromatic aberration under control. As a consequence of the growing desire for larger apertures, the length of the telescope increased dramatically, such as with the range of long and unwieldy instruments constructed by Hevelius.

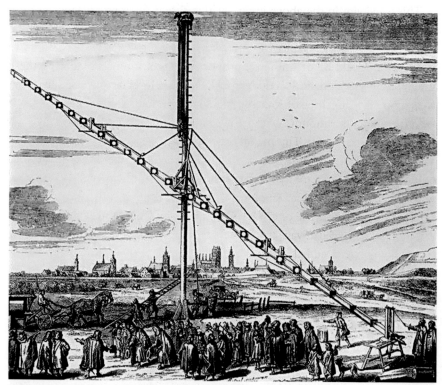

This illustration from his publication *Machina coelestis* (first part, 1673), which contained a description of his instruments, shows the 150-foot (45-metre) Keplerian refractor built by Johannes Hevelius and which stood on the shore of the Baltic Sea near his home town of Danzig (now Gdańsk). (Wikimedia Commons)

This brings me on to the way these telescopes were mounted. It is difficult to even imagine building and using something of this magnitude. The number of people that would have to be involved just to use it would have been immense. Problems that would have had to be overcome included deciding on the best lightweight materials to use and how many lenses would have to be employed to correct the image, as well as what distance apart they would need to be and where to space them so as to keep all things in balance.

The whole mechanism would have been set up in a way comparable to a giant set of weighing scales, with a pivotal shaft and series of block and tackle systems to slew from one target to the next using a mob of workers and sheer horse power.

This is evident in the above image, which shows the series of pegs used to raise and lower the optical tube assembly of Hevelius' giant refractor. As the positions changed the balance of the instrument would shift. The resulting flex in the tube would be of immense proportions, pretty much rendering these grand designs redundant. It is understandable that Hevelius only used his great Keplerian refractor three of four times before abandoning the design.

In the early years mounts were purely altazimuth in design, and usually with no sort of clutch system. This meant that the telescope would pivot on a central axis, resulting in there being no way of keeping the telescope on target without holding it still. This meant that if the observer had an unsteady hand, their view would be hindered and focusing made almost impossible.

This brought in the use of different materials to compensate, such as copper and brass rings placed at various intervals along the tube assembly. Various new components were added such as weighted lens caps to act as small counterweights attached via chains to help balance the weight of either objective lens or the focuser. As times moved on new mechanisms were developed to stable the telescope more accurately, an example being the introduction of a primitive clutch system, such as the basic locking nut still used today on basic mounts. This would enable the user to lock off the axis and stabilize the scope. However, with the lengths of some of the refractors of this time, there was still flexing in the optical assembly causing chromatic aberration.

Different styles of mount were soon developed, such as

This picture, taken from Christoph Scheiner's book *Rosa Ursina* (1673), shows Scheiner's Helioscope, thought to be the earliest example of an equatorial mount. (Wikimedia Commons)

Scheiner's Helioscope the first telescope mounted in such a way as to compensate for the rotation of the night sky. Constructed by Christoph Scheiner to assist his sunspot observations, this was a specially-designed telescopic solar projection instrument offering a definite advancement in mount design.

The development of mountings moved on quite rapidly during the following years, with the introduction of various new technologies, such as gearing and slow motion drives. This led to the production of the German equatorial mount, first developed by Joseph von Fraunhofer for the Great Dorpat Refractor, built by Fraunhofer and completed in 1824. With a 4-metre (13.4-feet) focal length and an aperture of 9.6-inch (24 cm), this was the first achromatic refracting telescope, its revolutionary mounting making observation a much easier task.

The brass right ascension (RA) drive fitted to the telescope made this instrument the first in the world to automatically track objects across the sky. Driven by clock weights that were set prior to the observing run, the drive was

The Great Dorpat Refractor housed at the Old Tartu Observatory in Tartu, Estonia. (Terry Nakazono)

adjusted and set to keep the objects centred in the eyepiece for uninterrupted study. However, the RA drive lacked rigidity, due to the drive gear itself being too small and making the mount wobble. This is an engineering and cost trade-off that is made in the lower end German equatorial mount designs to this day.

The era of extremely large telescopes, offset counterbalance systems and clock weight drives was born, making the balancing of large refractors and other types of telescope a simple task. This paved the way for other variations of the equatorial mount, including the English or Yoke mount and the Cross-axis mount.

The Yoke system has a frame or "yoke" with right ascension axis bearings at the top and bottom ends and a tube assembly attached inside the midpoint of the yoke. This allows the telescope to swing on the declination axis. The tube assembly is usually fitted entirely inside the fork and there is no need for a counterweight system.

These systems have now been superseded by harmonic drive systems, the high torque ratios of which eliminate the need for balancing the telescope all together, giving highly precise tracking and guiding. The future of the telescope mount is advancing well, with new developments in ways of balancing and tracking. However, there is much to be said for looking through the eyepiece of an old wobbly vintage telescope. With perhaps many years of history behind it, the telescope you are peering through may have been used to discover a new celestial object or to target and observe an early celestial event.

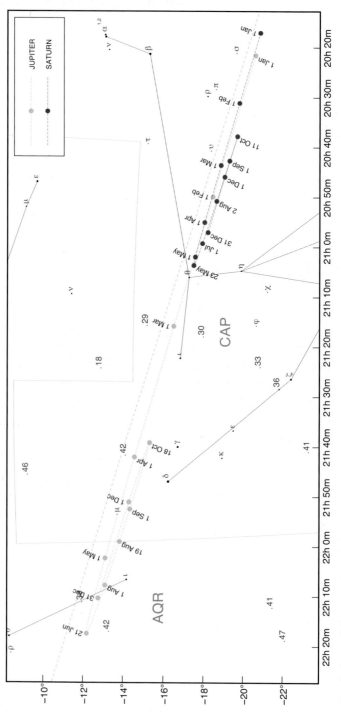

Jupiter and Saturn in 2021

August

New Moon: 8 August
Full Moon: 22 August

MERCURY is at superior conjunction on the first day of August and switches from the morning to evening skies where it remains for the next two months. This apparition is virtually undetectable to observers in northern temperate latitudes due to the low altitude of the planet, but provides this year's best evening views of Mercury for astronomers in the southern hemisphere. Mercury and Mars are at conjunction on 19 August, with Mercury found just 0.1° south of Mars. Mercury is brightest at the beginning of this apparition but is still a zero-magnitude object by the end of the month.

VENUS is the −4.0 magnitude evening star, climbing ever higher into the western skies of the southern hemisphere and equatorial regions. It remains disappointingly low to the horizon for northern temperate zone viewers, however. When viewed through a telescope, the waning gibbous planet grows in apparent diameter over the course of the month, from 12.8 arc-seconds to 15.2 arc-seconds.

EARTH is treated to a fine display of the Perseid meteor shower this month, the details of which may be found in the section entitled *Meteor Showers in 2021*. The Full Moon of 22 August is the third of four during the current astronomical season, making it a 'Blue Moon' by the original definition. On the penultimate day of the month, the Last Quarter Moon is at its nearest apogee of the year.

MARS appears in the evening sky in the constellation of Leo, setting soon after the Sun, and may be easier to view from southern latitudes than from the north. It is a second-magnitude object, about as dim as it will get as solar conjunction approaches. Mercury appears only 0.1° south of Mars on 19 August; although Mercury is the smaller planet, it is much brighter at magnitude −0.5.

JUPITER arrives at opposition on 19 August, a brilliant −2.9 object on the border of Aquarius and Capricornus. As viewed through a telescope, the gas giant appears 49.1 arc-seconds wide. Jupiter is visible throughout the night and is best viewed from the southern hemisphere where the ecliptic rises high overhead.

SATURN is at opposition on the second day of the month and is visible all night in Capricornus. Shining brightly at magnitude +0.2, the planet's disk has an apparent diameter of 18.6 arc-seconds. Add in the rings and you get an object 42.3 arc-seconds wide. The rings are tilted at an angle of 18.2° relative to Earth and are still widening slightly from their minimum value in May. The 26th of the month marks the 40th anniversary of the **Voyager 2** flyby of Saturn.

URANUS is found shining at magnitude +5.8 in the constellation of Aries. The waning crescent Moon is 1.7° south of the planet on the first day of the month and the distant planet reaches west quadrature six days later. Uranus reaches its maximum declination north (+15.8°) for the year on 19 August; the following day it arrives at a stationary point and goes into retrograde motion. The waning gibbous Moon makes another closer pass of Uranus on 28 August. The green ice giant now rises well before midnight and is seen to best advantage in the dark skies of the southern hemisphere winter.

NEPTUNE continues its lengthy retrograde loop through Aquarius this month and is brightening slightly, to magnitude +7.8, as it approaches opposition next month. The faint planet is best viewed from southern latitudes where it rises early in the evening and soars high overhead on the ecliptic. However, northern hemisphere observers now have a much better chance of seeing this dim object than they have had earlier this spring and summer.

Spare a Thought for the Engineers

Rod Hine

We are all quite blasé about the data rates of our internet connections. Most internet content downloads in the blink of an eye and whole movies can be imported in minutes or less, such is the blistering speed achieved with fibre optic cables. So, it's a sobering thought that New Horizons sent data back to Earth from its encounter with Ultima Thule (now officially known as 486958 Arrokoth) at a mere 1,000 bits per second, much slower even than the dial-up connections of twenty years ago. According to New Horizons Principal Investigator Alan Stern, speaking in December 2018, it will have taken 20 months to return all the data from the fly-by on New Year's Day. Because the probe had to swivel during the encounter to make best use of the imaging instruments it was some hours before controllers on the ground received confirmation of the success of the mission and it took some time to see the first pictures.

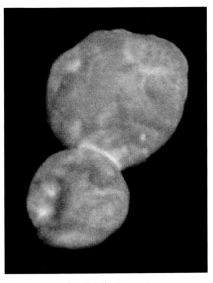

This image, taken by the Long-Range Reconnaissance Imager (LORRI), is the most detailed of Ultima Thule returned so far (at time of writing in June 2019) by the New Horizons spacecraft. It was taken at 5:01 Universal Time on 1 January 2019, just 30 minutes before closest approach, from a range of 18,000 miles (28,000 kilometres), with an original scale of 459 feet (140 meters) per pixel. (NASA/Johns Hopkins University Applied Physics Laboratory/Southwest Research Institute)

Communication over such huge distances by New Horizons and the Voyager probes is only possible because of the power of digital processing of the incredibly weak radio signals received on Earth by the antennas of NASA's Deep Space Network. It was the work of the brilliant but reclusive digital pioneer Claude Shannon which laid the foundations for the technology that enables this system to work at all, and which underpins every aspect of our modern information age. Shannon published *A Mathematical Theory of Communication* in the *Bell System Technical Journal* in 1948 and showed that theoretically it is possible to send data perfectly without errors at an arbitrarily low signal-to-noise ratio given suitable encoding and decoding techniques and an appropriate data rate. The author still has a well-thumbed copy of this paper, commonly known as "The Magna Carta of the Information Age", from his university studies in 1966!

The numerous publications of American engineer, mathematician and inventor Claude Shannon pioneered many concepts vital to the information age, always with a sound mathematical basis, allowing electrical and electronic engineering to progress from the ad-hoc methodology of early pioneers such as Bell, Edison and Marconi to the rigorous designs for high performance systems that are now commonplace. (Konrad Jacobs, Erlangen / MFO / Wikimedia Commons)

Although Shannon was the first to set out clearly the mathematical basis for communication, even establishing the link between information and entropy, some forms of storing and encoding data in a recognisably digital manner date back to the 18th century in the form of paper tapes used to control looms. The famous Jacquard loom, developed in 1804, used strings of punched cards to weave patterned fabrics and carpets. In 1842, the Frenchman Claude Seytre patented the use of punched paper rolls in "self-playing" pianos and by the middle of the 19th century the telegraph system used punched tape to send data automatically. Telegraph technology gradually evolved from a simple five-bit code patented in 1874 by the French telegraph engineer Jean-Maurice-Émile Baudot (after whom the Baud rate is named) to 7- and 8-bit codes. To detect errors in the received signal, an extra bit called the parity bit is added according to whether an odd or even number of holes has been punched for the actual data. Any single error in the received data would be obvious because the parity bit would not match so the receiving station could ask for the data to be retransmitted.

This was an effective but very inefficient way to get error-free data so attention then turned to more sophisticated encoding schemes that could both detect and correct errors without the need to re-transmit. The post-war rise of the electronic computer and the invention of the transistor in 1947 gave engineers the processing power to put Shannon's insight to good effect and by the 1950s, data could be transmitted at great rates by radio and rapidly stored and retrieved on magnetic tapes, disks and drums with perfect accuracy.

Engraving of Jean-Maurice-Émile Baudot by Antonin Delzers, (Antonin Delzers/ Wikimedia Commons)

By the time that the Voyagers were designed in the early 1970s, the best available error detecting and correcting codes were those introduced in 1960 by American mathematicians Irving Stoy Reed and Gustave Solomon (the Reed-Solomon codes), and it is these coding schemes that still enable the reception of data from the edge of interstellar space. The raw data is encoded into blocks and sent along with extra data generated from each block's contents. At the receiver the extra data is used to detect and correct any errors that have crept in due to noise. A very similar system allows CDs and DVDs to work perfectly despite scratches on their surfaces and keeps our modern digital TV pictures pixel-perfect. Only when the noise or disturbance exceeds the limiting design parameters does the signal show any degradation at all.

In the case of the Voyagers the transmitted signal is 22 watts, of which around 10^{-18} watts is received by a typical dish, with a path loss of some 31 orders of magnitude before the antenna gain is taken into account. The data rate is a mere 160 bits per second but so valuable is the data from deep space that extra computer processing is done on any corrupted data to try to reconstruct the signal, a process that is not feasible for domestic purposes.

The 70 metre Antenna at Goldstone, California. Part of the Deep Space Network, it was built in 1963, enlarged in 1988 and is still used on a daily basis to communicate with space missions. (NASA)

September

New Moon: 7 September
Full Moon: 20 September

MERCURY is well-placed in the evening sky for southern hemisphere observers but is too close to the western horizon to be easily visible from northern temperate latitudes. The planet continues to dim over the course of the month, beginning at zero magnitude and ending at +1.6. At 26.8°, this month's greatest elongation east (on 14 September) is the largest of the year. A week after this event, Mercury passes 1.2° south of Alpha (α) Virginis or Spica. Mercury reaches a stationary point and goes into retrograde on 27 September.

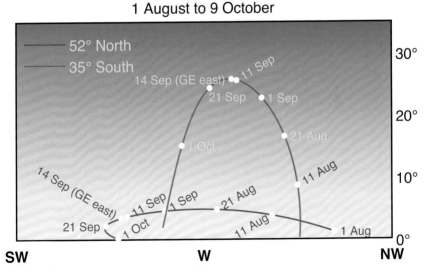

Evening Apparition of Mercury
1 August to 9 October

VENUS passes 1.4° north of Spica on the fifth day of the month. The evening star is slowly brightening, from magnitude −4.0 to −4.2, as its apparent diameter grows from 15.2 arc-seconds to 19.0 arc-seconds and its distance from Earth decreases to under 1 au. Venus is best viewed from the southern hemisphere and equatorial regions; it remains a disappointment at more northerly latitudes.

EARTH reaches another equinox on 22 September, marking the instant when the Sun crosses the celestial equator north to south. This ushers in spring for the southern hemisphere and brings on autumn in the north. The crescent Moon is 2.9° north of Praesepe, the open star cluster in Cancer, on the fourth day of the month, and about the same distance south of Pollux, the brightest star in the constellation of Gemini, on the last day. The Full Moon of 20 September is popularly called the 'Harvest Moon'.

MARS is low in the west at sunset, buried deep in the light twilight and difficult to spot at magnitude +1.8. On the fifth day of the month, it moves from Leo to Virgo, and is soon unobservable as it approaches conjunction with the Sun early next month. The third day of September marks the 45th anniversary of the arrival of the **Viking 2** lander on Mars, two months after **Viking 1** in July.

JUPITER is located in Capricornus, still in retrograde and still best seen from southern latitudes. At magnitude −2.7, it is far brighter than nearby Saturn. Jupiter is visible as soon as the skies turn dark and does not set until well after midnight.

SATURN is now past opposition and is an evening sky object, best viewed from the southern hemisphere where the constellation Capricornus rises high in the sky. Although the planet is getting farther from Earth, its rings are continuing to open up slightly so Saturn remains at around magnitude +0.2.

URANUS is only 1.2° north of the waning gibbous Moon on 24 September but the brightness and proximity of our satellite will make the sixth-magnitude planet impossible to see with the naked eye. Found in the zodiacal constellation of Aries, Uranus rises mid-evening and is well-placed for observation by midnight.

NEPTUNE arrives at opposition on 14 September, meaning it rises at sunset and sets at sunrise. It shines at magnitude +7.8 in Aquarius and appears as a blue disk just 2.5 arc-seconds in diameter in a small telescope.

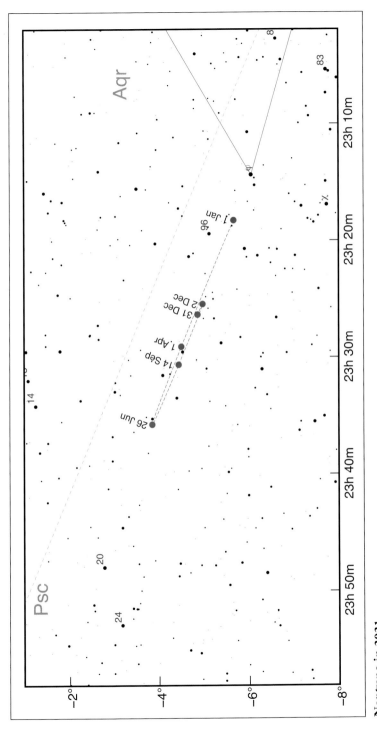

Neptune in 2021

Burying the Sun

Carolyn Kennett

Dusk is just an illusion because the Sun is either above the horizon or below it. Nicholas Sparks.

As a young child the theoretical physicist Richard Feynman would often walk in the woods with his father and he reminisces how his father would spend this time teaching him about science. In his lecture *What is Science?* Feynman said "I learned then what science was about: it was patience. If you looked, and you watched, and you paid attention, you got a great reward from it–although possibly not every time." In a similar manner humans have watched the world about them, in particular the sky and recorded the movements within it since the earliest of times. Sometimes their efforts may have been thwarted by clouds, but with patience and time, an understanding of the motions was built up.

Central to their father–son discussion was the role of the Sun and its paramount impact on the world. The importance of the Sun has been recognised throughout history. This luminous object defines the lives of people it shines on, and the clocklike regularity of its rise and set positions were well understood by early humanity. Due in part to an understanding of the Sun's significance, symbolic representation of the Sun stretches far back into prehistory. Early designs including the Sun are not uncommon and by the Bronze Age examples of solar symbolism are suggested across a range of medium including metals, stone and ceramics. Drawings and designs were often abstract in nature and representations included circles, waves and cross shapes. The cruciform shape in particular has been linked to the Sun by Mary Cahill and her work on Irish Sun discs.[1] Flat circular objects made of gold, they were designed to shine brightly when sunlight hits them. The etching of a cross on the surface shows the rays

---

1. 'Here comes the Sun – solar symbolism in Early Bronze Age Ireland', Mary Cahill, *Archaeology Ireland* 29(1), 26-33, 2015.

of the Sun in a conceptual way, maybe representing various solar events, such as sundogs, pillars, rays or halos.

Some objects include this cruciform design but in a position which was not so readily available to see. This design was also put on the interior of a number of ceramic urns. Although perhaps not as glamorous as on a golden disc these urns incorporate a similar design. One set of examples include the Trevisker style urns from the south west of England. Dating from the Bronze Age (approximately 1,500 BC) the style is known for its hash-dash and zigzag lines and use of local materials such as gabbroic clay from the Lizard peninsular. Many of these urns were utilised for a variety of mundane jobs but some examples would ultimately be used as burial objects and include the cremated remains of the dead.

Tregaseal Tomb, burial location of a crossed based urn. (Carolyn Kennett)

One such urn was found in the Tregeseal chambered tomb. It was recovered by archaeologist William Copeland Borlase in 1879 and now resides within the British Museum. Found at the end of the passage, in a separate enclosed area it was recovered almost complete by Borlase during his excavation. The Tregeseal urn is a large example of 21 inches in height and includes a cruciform shape in the interior of the base. It is worth saying that the cross itself would not be integral to the structure of the urn. There seems to be no practical explanation to why an urn of this size would have this addition in the base. It is therefore interesting to consider if some Trevisker urns could be following the tradition

The abstract decorations on gold lunula like the Blessington Lunula pictured here were made so sunlight optimised the designs, highlighting the close connection to the Sun that these people had. (Wikimedia Commons/John Maynard Friedman)

of Sun discs and offer a design which is indicative of solar symbolism? One aspect which strengthens the case of this idea is that the urn would have started as a disc shape on which the base cross design would be added. After a circular disc shaped base was made the sides would have been added to the urn creating the final vessel. Looking down from the mouth of the urn it would have clearly taken on representation of the Sun in a similar way to the Sun disc design.

The Tregeseal example was discovered within the tomb, rear ended - base facing the ceiling and inside were a number of cremated remains. In this final resting place the cross would have been above the remains which could have been intentional as many funerary urns are found base up. A cross at the top could represent a number of ideas. A Sun in the dark? A route to the heavens? A set Sun? Added to this the orientation of the chamber itself meant that it would only have been illuminated on the winter solstice sunrise. This orientation towards the solar extreme of winter solstice sunrise is in common with other local tombs at Bosiliack and Pennance. At Tregeseal on the winter solstice sunrise the rays of sunlight would have shone down the passageway hitting the back upright stone at the rear. The urn was positioned behind this back stone in a separate chamber or cist. Currently it is difficult to understand the full design of the tomb, as this rear section has been removed and we are reliant on the original plan from the Victorian excavation led by Borlase. It is worth considering if the positioning of the urn in a separate section was intentional. This position would ensure that no sunlight would reach the urn at any time. The symbolism may show that the urn was providing its own sunlight, even on this important day of renewal in the solar calendar – winter solstice.

The final resting place of this urn came most likely after a life of servitude. A similar Trevisker urn found in Kent had traces of animal fats within it and had been used as a transportation vessel for food before its final role as a funerary urn. Trevisker style urns are not the only vessels to include crosses on their base. Although in general this cruciform addition to urns is rare and seems to be reserved for the more decorative funerary examples. Other examples of cross based urns have been found in Ireland, Scotland and Yorkshire.

As a final thought I doubt we shall ever know the full truth behind this oddity in the design, but it is nice to consider an idea that the cross was an addition by our ancestors to bring sunlight into the darkest of places, and in effect they were burying the Sun with their dead.

October

New Moon: 6 October
Full Moon: 20 October

MERCURY departs from the evening sky early in the month, arriving at inferior conjunction on 9 October. It is less than 3° away from Mars on the same day but neither event is visible due to the proximity of the Sun. Direct motion recommences on 18 October and Mercury attains greatest elongation west (18.4°) just one week later. This apparition of Mercury is poor for southern hemisphere astronomers but it is the best morning appearance for observers in northern temperate latitudes, with the tiny planet climbing to 15° or more in altitude around the time of greatest elongation west on the 25th.

Morning Apparition of Mercury
9 October to 29 November

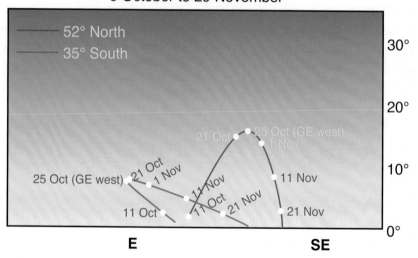

VENUS is just under 3° south of the waxing crescent Moon on 9 October but has a closer encounter with Antares (α Scorpii) a week later when the two celestial bodies approach to within 1.4° of each other. On 29 October, Venus attains a greatest elongation east of 47.0°. When viewed through a telescope, Venus appears almost half lit and it increases in apparent diameter from 19.0 arc-seconds to 25.8 arc-seconds over the course of the month. It also continues to brighten, shining at magnitude −4.4 by the end of October. The evening star is now high above the western horizon when viewed from the southern hemisphere and the tropics but hovers around 10° for astronomers in northern temperate latitudes.

EARTH experiences two major meteor showers this month, the Draconids in the first week and the Orionids in the latter half of the month, as well as the Southern Taurids, a minor shower peaking just after the Draconids. Observing details of these three events may be found in *Meteor Showers in 2021*. The waning gibbous Moon passes 2.6° south of Pollux (β Geminorum) on 27 October.

MARS is at solar conjunction on 8 October and has a close encounter with Mercury the following day. Neither of these events is observable and Mars remains hidden near the Sun for the entire month.

JUPITER reverses course on 18 October, going from retrograde to direct motion. The bright planet is at magnitude −2.6 and easily outshines Saturn with which it currently shares the constellation of Capricornus. An evening sky object, Jupiter sets around midnight for northern-based astronomers and a little later for planet watchers in the southern hemisphere.

SATURN reaches a stationary point on 11 October and resumes direct motion across the background stars of Capricornus. The rings reach a maximum opening tilt of 19.4° on the same day and then begin closing up again. Visible in the evening sky, Saturn sets before midnight from northern temperate latitudes and just after midnight from southern hemisphere vantage points. It is at magnitude +0.6 when it arrives at east quadrature on the penultimate day of the month.

URANUS and the Moon have another close encounter this month, with the nearly Full Moon passing just over a degree south of the faint planet on 21 October. However, the glare of the Moon will drown out any view of the magnitude +5.7 object. Uranus rises not long after sunset as it approaches opposition early next month. Look for it late in the evening in the constellation of Aries, near the sixth-magnitude blue-white star Omicron (o) Arietis (closest approach is 0.2° on 12 October).

NEPTUNE is now past opposition. It is already aloft in Aquarius during evening twilight and doesn't set until the early morning hours. The eighth-magnitude planet is best seen from northern latitudes where it remains above the horizon longer than for southern observers.

Tinkering with Time
The British Standard Time Experiment at 50

David Harper

The last day of October 2021 is a Sunday, so it marks the end of British Summer Time. The clocks go back to GMT, and everyone gets an "extra" hour in bed on Sunday morning. At a stroke, however, the evenings are darker, because sunset is an hour earlier by the clock. On Monday 1 November, most people will find themselves coming home from work in the dark for the first time since March. Every year at this time, there are calls to abandon the twice-yearly ritual of changing the clocks, and to keep British Summer Time all year round.

The last day of October is also the fiftieth anniversary of the end of a three-year period when Britain did exactly that. From 18 February 1968 until 31 October 1971, clocks were kept permanently an hour ahead of GMT in the British Standard Time experiment.

British Summer Time was introduced in 1916 as a wartime measure, and it was kept after the end of World War 1 because people found that the lighter evenings offered greater opportunities for leisure activities. Until the early 1960s (except for World War 2), it was kept for six months each year, typically from early April to early October. In 1961, due to public pressure, it was extended to seven months.

A sunset seen from the west coast of Portugal in October 2007. When the clocks go back in October, sunset is suddenly an hour earlier. (Wikimedia Commons/Joaquim Alves Gaspar)

Campaigners argued that keeping the clocks an hour ahead of GMT permanently would have many benefits. Lighter evenings would encourage outdoors activities all year round and thereby promote good health. Energy usage would be reduced, as people would not need to turn on heating or lighting until later in the day. In the autumn and winter months, children would be able to walk home from school in daylight, reducing the number of accidents on the roads.

To test these claims, Parliament passed the British Standard Time Act in 1968. In October of that year, the clocks remained an hour ahead of GMT. The Act provided for a review after two years, and in October 1970, an 84-page report was presented to Parliament.

According to the report, opinion surveys carried out in the winter of 1969/70 showed that the public in England and Wales were largely indifferent to British Standard Time, with only 15% of respondents claiming that they had suffered anything more than minor inconvenience from its introduction. The figure was a little higher in Scotland, but not significantly so. The main complaint, particularly in Scotland, was that people disliked getting up and going to work in the darker mornings in winter. It is worth noting that under British Standard Time, sunrise in London was after 9 a.m. from mid December to mid January, whilst in Edinburgh, the Sun didn't rise until after 9:30 a.m. for an even longer period.

The dark mornings were especially unpopular with postmen, who had to make their morning rounds in darkness for most of the late autumn and winter months, and with dairy farmers, who had to get their milk to collection points in time for the morning milk deliveries, which were dictated by their customers' need for fresh milk at breakfast. There was also concern in the construction industry, not only because builders would be starting work in the dark, but also because ice and frost would not have had time to melt, making building sites even more hazardous at the start of the working day.

Road accidents did decrease in the early evening period, but more accidents were recorded in the morning between 8 a.m. and 9 a.m. Among children walking or cycling to school, the increase in morning accidents outweighed the decrease in the early evening. Critics also noted that the decrease in evening accidents might be explained by the introduction of a strict drink-driving law in 1967, which included mandatory breathalyser testing.

In the Commons debate in December 1970, many MPs spoke strongly against continuing the experiment, especially those representing Scottish constituencies. They cited opposition from their constituents, as well as concerns about child safety and the welfare of postmen and other early-morning workers. The Government allowed a free vote on the issue, and the House voted by 388 to 91 to end the experiment. On 31 October 1971, Britain went back to GMT and the twice-yearly clock change.

The United States also experimented with year-round daylight saving time (DST) during the winter of 1974–5 as a response to the 1973 oil crisis, but went back to observing DST only in the summer from 1975 onwards. The U.S. Congress has extended the period of DST twice since then. As in Britain, energy saving and road safety have been cited as arguments in favour. The leisure industry also supported it. The votes of the two senators from the state of Idaho were secured for the 1986 extension of daylight saving time when it was pointed out that longer evenings offered more opportunities for the sale

Idaho potatoes swayed the state's senators to vote to extend daylight saving time in 1986. They are also celebrated on Idaho licence plates. (Wikimedia Commons)

of fast food. Idaho is famous as a potato-growing state, and the Idaho Russet potato is especially good for making French fries.

There have been numerous attempts to bring back permanent British Summer Time, or to move Britain to the same time zone as France, Germany and other western European countries. Between 1994 and 2011, eight bills were introduced in Parliament, but none had government support. At the time of writing, there are no plans to repeat the British Standard Time experiment of 1968–71. In the United States, the idea seems to be gaining traction. Seven state legislatures have already passed bills supporting permanent daylight saving time, and a further 29 states are considering it. However, federal law currently does not allow states to implement permanent DST unilaterally.

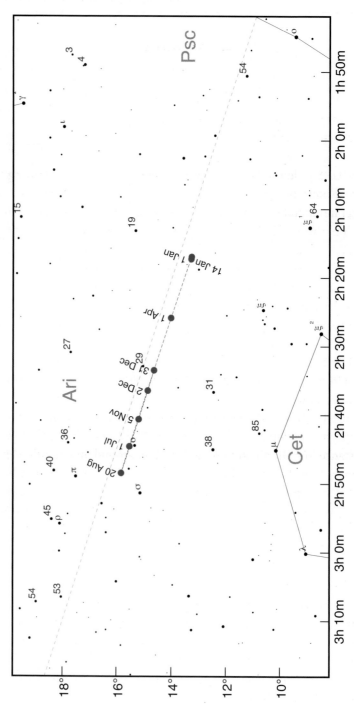

Uranus in 2021

November

New Moon: 4 November
Full Moon: 19 November

MERCURY is descending back toward the eastern horizon, drawing to a close the best morning apparition of the tiny orb for planet watchers in northern temperate latitudes. The planet is slowly brightening over the course of the month but will be lost to view in the dawn skies from around mid-November. Any inhabitants in the far northern reaches of Canada may see the very old crescent Moon occult Mercury on the third day of the month beginning around 18:30 UT. Mercury and Mars have another close encounter this month, on 10 November, when the two planets are a degree apart but only 11° from the Sun. Mercury undergoes its final superior conjunction of the year on 29 November.

VENUS is occulted by the waxing crescent Moon on 8 November in an event visible from north eastern Asia from approximately 05:30 UT. The evening star continues to brighten, from magnitude −4.4 to −4.7 over the course of the month, and takes on a waning crescent shape when viewed through a telescope. Its apparent diameter increases from 25.8 arc-seconds to 39.1 arc-seconds and its distance from Earth drops to under 0.5 au around mid-month. Venus begins to slowly lose altitude night by night as seen from the southern hemisphere but that's still the best place to observe this bright planet. It actually begins to climb a bit above the western horizon for astronomers in northern temperate latitudes but remains very low for the remainder of the year.

EARTH experiences a partial lunar eclipse on 19 November, the details of which may be found in *Eclipses in 2021*. Two noteworthy meteor showers also take place this month, the Northern Taurids (first half of the month) and the Leonids (second half of the month). Consult the section *Meteor Showers in 2021* for more information. The crescent Moon occults the two inferior planets this month, with Mercury vanishing first on 3 November and Venus disappearing five days later. The waning gibbous Moon is found 2.5° south of first-magnitude Pollux (β Geminorum) on 24 November.

MARS is now a morning sky object but remains a difficult target as it rises in dawn skies this month. The very old crescent Moon pays a visit on 4 November and Mercury closes to within a degree of the red planet on 10 November but the latter event takes place only 11° degrees away from the Sun. Northern hemisphere observers have the best chance of seeing the planetary conjunction before sunrise. Mars crosses the border from Virgo into Libra on 11 November.

JUPITER is the brighter of the two gas giants currently residing in Capricornus, blazing away at magnitude −2.4. An evening sky object, Jupiter sets before midnight. It reaches east quadrature on 15 November.

SATURN shines at magnitude +0.6 in Capricornus this month. It is an evening sky object, already aloft at sunset and vanishing in the west before midnight. The tilt of the rings is closing and the planet is looking smaller in telescopes as the planet slowly draws away from Earth following August's opposition.

URANUS reaches opposition on 5 November. At magnitude +5.7, it is barely visible to the naked eye. In a telescope, it presents a greenish orb 3.7 arc-seconds in diameter. The Moon has been making a series of increasingly close approaches to the planet but now begins to pull away. The two celestial bodies appear 1.3° apart on 18 November. Uranus is in Aries, rising at sunset and setting at sunrise.

NEPTUNE continues its retrograde path through Aquarius this month, shining at magnitude +7.8. It is visible in the evening sky and is best viewed from the northern hemisphere where the nights are long and dark. It sets around midnight by the end of November.

Your Name in Space

Peter Rea

The idea began with the launches of Pioneer 10 (1972) and Pioneer 11 (1973) bound for Jupiter. As both Pioneers would be the first spacecraft to be sent on a path that would carry them out of the Solar System, they each carried an identical commemorative plaque featuring a pictorial message which showed details about us. The plaque depicts naked figures of a human male and female standing in front of the spacecraft for scale. It further shows a number of symbols designed to provide information about the origin of the spacecraft. These include one showing the Pioneer craft leaving the third planet from our Sun and heading out into space after passing the fifth planet, as well as a diagram showing the relative position of our Sun in relation to 14 known pulsars. Whilst

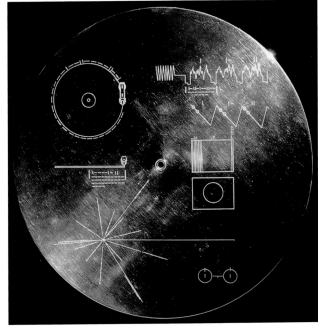

This side of the Voyager gold record shows the location of our Sun in relation to known pulsars and instructions on how to play the disc. Honestly! (NASA/JPL-Caltech)

Shown here are Cassini science and mission design manager Charley Kohlhase on the left and Cassini program manager Richard Spehalski holding the DVD before installation on the Cassini orbiter shown in the background. (NASA/JPL-Caltech)

not carrying any actual names into space, they do send a message to the stars with information as to who and where we are, despite the rather remote chance of them ever being found.

The idea of sending messages on a spacecraft was taken further with the launches of Voyagers 1 and 2 in 1977. Attached to the side of each of the Voyager spacecraft is a gold-plated copper disc similar in size to the vinyl records we use to play music. It contains spoken greetings from around the world in many languages. It also contains images and sounds of life on Earth and various pieces of music as diverse as Beethoven and Chuck Berry. Like a message in a bottle, it was sent out into the universe for anyone with the means to capture it and figure out how to play it. Instructions on how to do this were shown diagrammatically on one side of the record.

In 1997 the general public were given the chance to participate in the Cassini / Huygens mission to Saturn. Your name could be submitted via the mission website at the Jet Propulsion Laboratory of the California Institute of Technology (JPL). The DVD bearing 616,403 digitized signatures (including my own) from 81 countries was attached to the Cassini spacecraft shortly before launch on 15 October 1997. I was fortunate enough to see it launched.

The Cassini / Huygens mission was a joint mission between the USA and the European Space Agency (ESA). The Cassini Orbiter was built by the Jet Propulsion Laboratory (JPL) for the USA, with ESA providing a Titan atmospheric probe called Huygens that would be attached to Cassini. After arriving in orbit around Saturn, the probe would be released to parachute through the dense atmosphere of Titan and land on the surface. The DVD containing thousands of signatures was attached to the Huygens lander before launch, and that disc now resides forever on the surface of Titan.

There have been many more opportunities over the last two decades. The earliest evidence I can find of names going on a planetary spacecraft is when the Planetary Society had a disc with 100,000 members attached to the Russian Mars 96 mission. However, this mission did not make it to Mars after a launch failure. Since then many Mars missions have given the general public opportunities to fly their names to Mars, including Mars Pathfinder 1997, the Mars Exploration Rovers Spirit and Opportunity 2003, Mars Phoenix 2007, Mars Science Laboratory 2011, Mars Maven 2013 and the InSight mission to Mars which launched in 2018 with over 2 million names. I took advantage of all of these, and at the time of writing I have just submitted my name to be

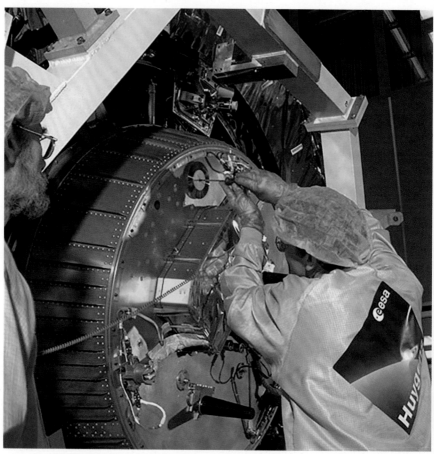

The CD with digitised names being attached to the upper side of the Huygens probe. (ESA)

attached to the NASA Mars Rover 2020. A quick glance at the mission website tells me the number of people submitting names has already passed 2 million and it has only been going for a few weeks.

Whilst Mars was a popular destination, it was not the only one. Launched on 18 June 2009, the Lunar Reconnaissance Orbiter also participated in sending 1.6 million names on a microchip to the Moon. Other missions to solar system objects proved equally popular. The following spacecraft had names onboard. The Deep Impact probe was launched toward Comet Tempel 1 in January 2005, a disc containing 650,000 names being mounted on the impactor portion of the spacecraft. When the impactor slammed into Tempel 1 on 4 July 2005, it and

The micro chip with over 2 million names being attached to the upper deck of the InSight mission to Mars. (NASA/JPL-Caltech/Lockheed Martin)

all the names it carried were vaporized. The Stardust spacecraft, launched in February 1999 toward the comet Wild 2, carried over a million names. More recently the Parker Solar Probe, launched on 12 August 2018, carried 1.1 million names on its circuitous journey around the Sun. By late-2025 the orbit of the probe will take it to within just 7 million kilometres of the Sun.

Sending names alone, whilst allowing an individual to feel a part of the mission, could leave that person wondering whether they could do more. Well for three missions, at least, the answer was yes. The Mars Exploration Rover Opportunity landed on Mars in 2004. Opportunity greatly surpassed its expected duration on the red planet, having travelled more than 45 kilometres. In August 2012 the rover arrived on the rim of Endeavour crater which it explored for the next 6 years. Gradually working its way anticlockwise around the rim, it arrived at what would turn out to be its final location by early-2018. In June of that year a large dust storm encircled the planet, the effect being to coat the power generating solar panels with dust which greatly reduced the amount of electricity it could generate. This caused Opportunity to go into a low power mode to conserve what little power was being generated. JPL allowed the public to send a digital postcard to Opportunity giving it "encouragement" to wake up and start transmitting again back to Earth. Despite these postcards, Opportunity did not wake up and further attempts were cancelled in the spring of 2019.

JPL gave the public another chance to send a postcard to a Mars mission. The mission website for the Mars Science Laboratory Curiosity had postcard templates that could be filled in online. You just clicked in the template of choice and typed a short message to Curiosity. NASA's Deep Space Network Antennas did the rest, thereby presenting another chance for the planetary scientists of the future to feel connected to a mission.

Readers may recall that the New Horizons spacecraft successfully flew past Pluto/Charon in July 2015 and was later targeted to the Kuiper Belt Object, 2014 MU69, referred to at the time as Ultima Thule but now known by its official name of Arrokoth. The Johns Hopkins University Applied Physics Laboratory who controlled the spacecraft allowed participants to send a message via the mission website to New Horizons just before the flyby in January 2019. Although the number of words allowed was strictly limited, it gave the public (and me) a chance to feel connected to this very distant spacecraft.

Allowing the general public to send their names and occasional message to the planets and out of the solar system stimulates an interest in the space programme and encourages youngsters who could one day be the next generation of space scientists. We should not look on sending names to solar system spacecraft as a gimmick. It gives those who wish to participate the opportunity to connect to and feel part of a mission. If you should ever doubt this then recall the words of the Greek philosopher and writer Plutarch who said "The mind is not a vessel to be filled, but a fire to be kindled".

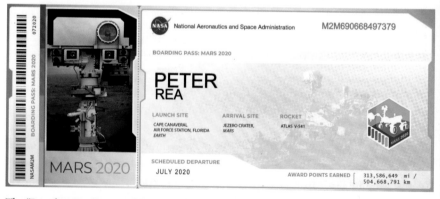

The "Boarding Pass" as proof that the authors name will be sent to Mars on the next NASA Mars Rover in 2020. (Peter Rea via JPL mission website)

December

New Moon: 4 December
Full Moon: 19 December

MERCURY reappears in the west at sunset around mid-month in an evening apparition which favours Mercury enthusiasts along the Earth's equator. Those lucky observers in the path of totality during the solar eclipse of 4 December may see magnitude −1.1 Mercury just 3° away from the eclipsed Sun. Mercury and Venus have one final get together for the year on 29 December when the two planets are around 4° apart.

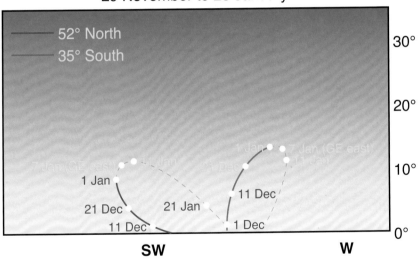

Evening Apparition of Mercury
29 November to 23 January

VENUS and the waxing crescent Moon are less than 2° apart on 7 December. Mercury joins Venus in the latter part of the month, with the two coming to within 4.2° (16° from the Sun). The evening star attains its maximum magnitude of −4.7 mid-month and is best seen from southern and tropical latitudes. However, look for Venus earlier rather than later in the month as the planet is losing altitude rapidly. Telescopically the planet presents as a large waning crescent, over 60 arc-seconds in diameter by the end of the year.

EARTH sees a total eclipse of the Sun on 4 December. This coincides with the Moon's nearest perigee of the year. Details of the eclipse may be found in *Eclipses in 2021* elsewhere in this volume. Two major meteor showers also take place in December, the ever-reliable Geminids mid-month and the Ursids a week later. Exact dates and observing conditions are explained in the section *Meteor Showers in 2021*. Solstice occurs on 21 December. The Sun reaches its most southerly declination on this day and marks the beginning of summer in the southern hemisphere and the onslaught of winter in the north. The Full Moon of December is this year's 'Micro Moon', the Full Moon with the smallest apparent diameter. The waning gibbous Moon makes another pass by Pollux, the lucida of Gemini, on 21 December, when it appears just 2.6° south of the bright star.

MARS is occulted by the waning crescent Moon twice this month. Early risers in northern and eastern Asia will have a chance to see Mars disappear behind the disk of our satellite from 23:00 UT on 2 December. On the last day of the month, observers in southern Australia will have a similar opportunity beginning around 18:30 UT. Now at magnitude +1.6, Mars begins the month in the constellation of Libra, moving briefly into Scorpius on 15 December and then into the non-zodiacal constellation of Ophiuchus on 24 December. The red planet passes through the descending node of its orbit on 19 December, moving north to south across the celestial equator.

JUPITER is visible as a magnitude −2.2 object in the western part of the sky at sunset, not vanishing from view until mid-evening or later. Jupiter begins the month in Capricornus but returns to Aquarius on 14 December.

SATURN sets in the early to mid-evening, depending upon latitude. Look for a magnitude +0.6 object in the vicinity of much dimmer star Theta (θ) Capricorni as soon as the skies darken.

URANUS remains in retrograde through the end of the year, slowly retracing its path through Aries. The waxing gibbous Moon passes it on 15 December. Uranus is now past opposition and is an evening sky object, setting just before dawn twilight.

NEPTUNE arrives at another stationary point on the second day of the month and resumes direct or prograde motion across the background stars of Aquarius. The blue ice giant is at east quadrature ten days later. Neptune is an evening sky object, setting around midnight at the beginning of the month and at approximately two hours earlier by the end of the year. At only magnitude +7.9, Neptune requires dark skies and a small telescope to reveal it.

Tycho Brahe

Paul Fellows

Even today we tend to try to divide astronomers into two groups, those concerned with the theory and the astrophysics of situations, and those who carry out the detailed observations necessary to back-up or refute those theories. The first group, without the efforts of the second to underpin and check their ideas, are really guessing, and would be able to come up with any number of notions about how the world works with no way to tell which might be correct. Thus it has ever been, even before the invention of the telescope, although it is often those scientists who propose new theories, such as Copernicus with the heliocentric view of the universe or Newton with his explanation of the laws of motion, which grab the limelight. A favourite example of mine of this is the interplay between a master and his student, in the persons of Tycho Brahe (the master) and his student Johannes Kepler. It was Kepler who got his name attached to the three laws of planetary motion which he proposed; although he had only been able reach his conclusions as a result of the exquisitely precise and detailed observational

This portrait of Tycho Brahe by Austrian artist Eduard Ender shows him as a very striking individual with his fiery red moustache – and an intact nose, which suggests that this image predates the famous duelling incident in which he lost it. (Wikimedia Commons / Eduard Ender)

A Question of Pronunciation

Tycho Brahe was born in Denmark on 14 December 1546 to parents who were part of the local aristocracy, and no doubt his somewhat privileged lifestyle set him on a course that lead to him becoming the most famous astronomer in Europe until his death in the year 1601. His name is often the subject of a debate as to how to pronounce it. The Anglo-Saxon tongue finds it a bit of a twister, and indeed the man himself seems to have changed it from the birth-name of 'Tyge (Ottensen) Brahe' to the more Latinized version that we use today. Pronouncing it in the revised form remains a problem and can vary with 'TIE' 'CO' or 'TEE' 'CO' for his first name and then 'BRA', with or without a 'HE', on the end. Personally I go for 'TIE' 'CO' 'BRA' 'HE' – but no doubt there will be others who say this is wrong; anyway, it is perhaps the least important thing about the man.

measurements of the positions and movements of the planets that were the life's work of his mentor.

Although his initial education in Latin and law at university set Tycho on a fairly traditional path for a young man of his situation, the observation of an eclipse triggered a fascination in him that was to define his life. He was fascinated by the idea that scientists were able to predicted eclipses years in advance and he began to study astronomy alongside his other work. The critical moment perhaps came when he discovered that star maps from the leading astronomers of the day did not agree with each other – clearly a major problem if you wished to make predictions of the movements of the Moon and planets relative to these. Tycho became convinced that new and better charts were necessary and set out on a course to compile the most accurate star charts possible. To this end he found it necessary to design and build new instruments for making precise positional determinations. These instruments were not telescopes (that invention had yet to come) but were devices for timing the passage of objects across the sky and determining their relative positions with far greater precision to any that the world had yet seen. He then used his measurements to produce extremely accurate tables of star and planetary positions. This data was Tycho's legacy to the world, and it was this with which Johannes Kepler was finally able to put to bed the idea of perfect circular motion of the heavenly bodies after he discovered that the paths of the planets were elliptical.

This engraving from *Astronomie Populaire* by French astronomer Camille Flammarion depicts Tycho observing the bright supernova of 1572. Now known as Tycho's Supernova, it was so bright as to be visible in daylight. (Wikimedia Commons / Camille Flammarion)

Tycho's work was not limited to positional astronomy. In 1572 he discovered a "new star" which appeared in the constellation Cassiopeia and which subsequently faded. Although today we would recognise this as a supernova, at the time it was a revolutionary discovery due to it flying in the face of the ideas of both Aristotle and the Church of a perfect and eternally unchanging view of the heavens. Modern observations show the expanding remains of what is now known as "Tycho's Supernova".

In 1577 his observations of a comet also revealed that it was more distant from the Earth than even the Moon, thereby putting paid to another Aristotelian notion which held that comets were atmospheric phenomena.

And of course, it was in 1600 that he hired the young astronomer and mathematician Johannes Kepler to be his assistant. Kepler carried out the work on the problem of predictions of the orbit of the planet Mars, and it

An image taken by NASA's Wide-field Infrared Survey Explorer (WISE) showing a number of interesting objects in Cassiopeia, including the expanding shell of gas from Tycho's Supernova (SN 1572), visible as a red circle towards the upper left of picture. (NASA/JPL-Caltech/UCLA)

was the results achieved by him that would ultimately lead to the death knell for another ancient idea – that of the perfection of heavenly motion with the Sun and planets in moving in perfect circles around the Earth. This is a slightly ironic twist in that Tycho himself was aware of the ideas of Copernicus but did not accept them, holding rather to the Earth-centric views. Tycho passed away just a year later and did not live to see the publication by Kepler of the solution to the problem that he himself has set.

Finally of course I come to the one fact that every story about Tycho is obliged to mention, which concerns his nose – or lack thereof. He was by all accounts somewhat hot headed and ended up fighting a duel in which his nose (or at least part of it) became a casualty, so leading to its replacement with a metal prosthesis. However, I would venture to say that progress generally tends to come from unreasonable people as a consequence of their drive and persistence, and that Tycho Brahe, the man with the golden nose, is no exception. It is pleasing that he is remembered each time we look at the Moon, in particular at the prominent impact crater Tycho which is named in his honour and which, located in the heavily cratered southern lunar highlands, plays host to the most extensive ray system on the lunar surface.

A mosaic of images from the Lunar Reconnaissance Orbiter showing the prominent lunar impact crater Tycho, which was named in honour of the famous Danish astronomer. (NASA)

Comets in 2021

Neil Norman

Comets are viewed as the highly unpredictable objects of the Solar System and can appear at any given moment with no prior warning. Although no object is expected to put on the show that the great Comet Hale-Bopp did in 1997 (this was one of the brightest comets seen for many decades and one of the most widely observed of the twentieth century), the next 'great comet' could be discovered at any time. It is therefore advisable to keep a close eye on social media or the BAA (British Astronomical Association) Comet Section page **www.ast.cam.ac.uk/~jds** where the latest comet discoveries are announced.

At the time of writing, a total of 89 comets are due to return to perihelion during 2021. These are made up of some 72 short period comets; eight objects that are suspected of being cometary in nature; five long-period comets; and the remaining four objects listed as being lost or defunct in activity. Although only four (see below) are expected to attain magnitude 10 or brighter, comets are famed for their unpredictability, and outbursts could occur leading to one or more of them becoming brighter at any time.

7P/Pons-Winnecke

Discovered in Leo by the great comet hunter Jean Louis Pons from Marseille, France on 12 June 1819, the comet that was to become known as 7P/Pons-Winnecke was lost until recovered by German astronomer Friedrich August Theodor Winnecke at Bonn on 9 March 1858. Winnecke was able to demonstrate that this comet and the one seen by Pons were indeed one and the same object, leading to it being given the name that we identify it with today.

The comet was viewed again in 1869 as predicted, although gravitational interactions

Friedrich August Theodor Winnecke. (Wikimedia Commons)

with Jupiter led to increases in its orbit occurring since the early part of the nineteenth century. 7P/Pons-Winnecke makes frequent passes close to Earth, such as in 1927 (0.04 AU), 1939 (0.11 AU), 1892 (0.12 AU), 1819 (0.13AU) and 1921 (0.14 AU) returns, and in 2021 it will approach to within 0.44 AU of our planet. However, although a fairly close pass, this will not result in the comet becoming particularly noticeable, with maximum brightness expected to be around tenth-magnitude.

Observers will be able to locate 7P/Pons-Winnecke as early as April, but when at its brightest, it will only be readily visible to observers in the southern hemisphere.

DATE	R.A.	DEC	MAG	CONSTELLATION
1 Apr 2021	17 53 20	+08 47 34	14	Ophiuchus
15 Apr 2021	18 38 08	+07 17 01	13	Ophiuchus
1 May 2021	19 33 37	+03 39 59	12	Aquilla
15 May 2021	20 26 03	+01 47 50	10	Aquilla

15P/Finlay

South African astronomer William Henry Finlay discovered this comet as an eleventh-magnitude object in Scorpius on 26 September 1886 while observing from the Royal Observatory, Cape of Good Hope. Over the following three months the comet attained ninth-magnitude with a recorded brightness of 8.5 as it passed closest to Earth around mid-December. The object was recovered by Finlay at the following return in 1893, again at eleventh-magnitude and reaching a maximum of magnitude 8.5 by late June. A close Earth approach at the 1906 return (0.27 AU) saw the brightness rise to magnitude 6. A close pass to the planet Jupiter (0.46 AU) in 1910 resulted in a slight increase in its orbital period, which led to the 1913 return being poor from the fact that perihelion took place when in conjunction with the Sun. No good returns were seen before that of 1953 and since then, nine have been seen.

With an absolute magnitude (i.e. if the comet was seen from 1 AU away from Earth) of just 10.3 15P/Finlay is a rather difficult object to observe at the best of times.

15P Finlay. 2015 Jan 18 at 0220UT. I-TEL T24 0.61 m f/6.5 + CCD. 5 minute exposure. FOV 10'. Tail in PA 66.(c) M Mattiazzo

Passing through Aquarius and lying at a distance of 1.3 AU from Earth, Comet 15P Finlay was in the early stages of outburst when this image was captured on 18 January 2015 from Swan Hill, Victoria, Australia. (Michael Mattiazzo)

DATE	R.A.	DEC	MAG	CONSTELLATION
15 Aug 2021	05 52 33	+25 57 32	10	Taurus
1 Sep 2021	06 50 18	+26 57 58	10.5	Gemini
15 Sep 2021	07 30 13	+26 55 57	10.5	Gemini
1 Oct 2021	08 07 20	+26 28 58	11	Cancer
15 Oct 2021	08 32 29	+26 04 49	11.5	Cancer
1 Nov 2021	08 53 35	+25 56 25	12	Cancer
15 Nov 2021	09 02 36	+26 18 56	12.5	Cancer
1 Dec 2021	09 02 41	+27 22 55	14	Cancer
15 Dec 2021	08 53 22	+28 45 25	16	Cancer

29P/Schwassmann-Wachmann

This comet was discovered by Arnold Schwassmann and Arno Arthur Wachmann at Hamburg Observatory in Bergedorf, Germany on photographic plates taken on 15 November 1927. The comet was at magnitude 13 and in outburst at the time of discovery. With a perihelion distance of 5.7 AU and aphelion point of 6.25 AU, and an eccentricity of just 0.0441, the orbit of 29P/Schwassmann-Wachmann is almost circular and lies entirely between the orbits of Jupiter and Saturn. The orbital period around the Sun is 14.65 years, and it will next reach perihelion on 31 October 2033.

Comet 29P/Schwassmann-Wachmann imaged on 16 October 2019 by Nick Haigh, Southampton, England. (Nick Haigh)

This object is known for its unpredictable outbursts, during which it can rise from its usual thirteenth-magnitude to around magnitude 9 in the course of a day. It has at least four very active regions upon its surface and rotates at a rather slow 12 days, meaning any outburst can be followed for long periods of time and studied in great detail. Needless to say, all reliable observations of 29P/Schwassmann-Wachmann are welcomed, and can be submitted to the BAA.

DATE	R.A.	DEC	MAG	CONSTELLATION
1 Jan 2021	02 14 06	+24 21 22	12 – may vary	Aries
1 Feb 2021	02 19 23	+23 49 55	12 – may vary	Aries
1 Mar 2021	02 32 45	+24 08 08	12 – may vary	Aries
1 Sep 2021	04 51 08	+31 08 55	12 – may vary	Auriga
1 Oct 2021	04 57 14	+31 53 00	12 – may vary	Auriga
1 Nov 2021	04 51 26	+32 15 36	12 – may vary	Auriga
1 Dec 2021	04 36 44	+32 58 06	12 – may vary	Perseus

67P/Churyumov-Gerasimenko

This now famous Jupiter-family comet was discovered by Soviet astronomers Klim Ivanovych Churyumov and Svetlana Ivanovna Gerasimenko on photographic plates taken in September 1969. The return of Churyumov–Gerasimenko in 2021 will mark the first return of the comet since its memorable rendezvous with the highly successful ESA Rosetta mission on 6 August 2014. It was then that Rosetta's lander Philae became the first spacecraft to land on a comet nucleus when it touched down on 12 November 2014.

The 2021 return could see the comet peak at ninth-magnitude in December as it approaches to within 0.41 AU of Earth. Best seen in the morning skies moving south later in the year, the comet will be visible in good binoculars from autumn onwards.

DATE	R.A.	DEC	MAG	CONSTELLATION
1 Aug 2021	01 53 15	+05 58 27	14	Pisces
15 Aug 2021	02 37 08	+09 55 10	14	Cetus
1 Sep 2021	03 42 34	+15 06 46	13.5	Taurus
15 Sep 2021	04 48 20	+19 11 55	13	Taurus
1 Oct 2021	06 13 43	+22 26 59	12.5	Gemini
15 Oct 2021	07 28 33	+23 15 32	12.5	Gemini
1 Nov 2021	08 46 45	+22 06 01	12	Gemini
15 Nov 2021	09 36 13	+20 26 26	11	Leo
1 Dec 2021	10 16 10	+18 49 21	10	Leo
15 Dec 2021	10 37 06	+18 11 13	9	Leo

Comet 67P Churyumov-Gerasimenko was in Libra and located at a distance of 3.1 AU when imaged on 14 November 2015 from Latisana, Italy. (Rolando Ligustri)

Minor Planets in 2021

Neil Norman

Minor planets is just another term for asteroids, the varying sized pieces of rock left over from the formation of the Solar System around 4.6 billion years ago. Millions of them exist, and to date around 794,000 have been seen and documented with nearly 500,000 having received permanent designations after having being observed on two or more occasions and their orbits now being known with high precision.

Most asteroids travel around the Sun in the main asteroid belt, which is located between the orbits of Mars and Jupiter. However, some asteroids have more elliptical orbits which allow them to interact with major planets, including the Earth. In all, there are around 2,000 which can approach to within a close distance of our planet. These objects are referred to as Potentially Hazardous Asteroids (PHAs). To qualify as a PHA these objects must have the capability to pass within 8 million kilometres (5 million miles) of Earth and to be over 100 metres across.

Objects of this size could pose a serious threat if on a collision course with Earth. It is estimated that several thousand exist with diameters of over 100 metres, and with around 150 of these being over a kilometre across. A

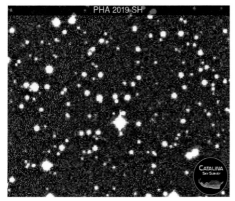

Discovery image of the Potentially Hazardous Asteroid 2019 SH. Discovered on 17 September 2019 by the Catalina Sky Survey, when at a distance of 0.11 AU from Earth, 2019 SH made its closest approach to our planet (0.05 AU) on 23 September 2019. Its next close approach to Earth will be on 23 March 2173 (0.06 AU). (Catalina Sky Survey / David Rankin)

large number of smaller asteroids, measuring anything between just a few meters in diameter to several tens of meters wide, pass close to our planet on a regular basis, with considerable numbers of smaller ones entering the Earth's atmosphere every day, burning up harmlessly as meteors.

Those observers with a particular interest in following these objects should go to the home page of the Minor Planet Center. It is their job to keep track of these objects and determine orbits for them. This page can be accessed by going to **www.minorplanetcenter.net** where you will find a table of newly discovered minor planets and Near Earth Objects (NEOs). At the top of the page is a search box

The discovery for which Giuseppe Piazzi is best remembered is that of the first minor planet Ceres. (Wikimedia Commons/JPL)

that you can use to locate information on any object that you are interested in, and from this you can obtain ephemerides of the chosen subject. The Minor Planet Center site is the one that all dedicated asteroid observers should consult on a regular basis.

The following minor planets are well placed for observation from northern latitudes during 2021 and are therefore ideal targets for backyard astronomers equipped with only moderate optical aid.

1 Ceres

With a diameter of 945 kilometres (587 miles) Ceres is the largest object in the asteroid belt. Discovered by Italian astronomer Giuseppe Piazzi from Palermo Observatory, Sicily on 1 January 1801, Ceres was initially believed to be a planet until the 1850s when it was reclassified as an asteroid following the discoveries of many other objects in similar orbits.

DATE	R.A.	DEC	MAG	CONSTELLATION
1 Aug 2021	03 56 54	+13 56 51	9.0	Taurus
15 Aug 2021	04 12 34	+14 42 17	8.9	Taurus
1 Sep 2021	04 28 32	+15 22 51	8.8	Taurus
15 Sep 2021	04 38 18	+15 46 68	8.6	Taurus
1 Oct 2021	04 44 36	+16 04 33	8.3	Taurus
15 Oct 2021	04 44 57	+16 16 13	8.0	Taurus
1 Nov 2021	04 38 16	+16 28 28	7.6	Taurus
15 Nov 2021	04 27 27	+16 39 06	7.3	Taurus
1 Dec 2021	04 11 49	+16 54 32	7.0	Taurus
15 Dec 2021	03 58 31	+17 14 04	7.4	Taurus

2 Pallas

Pallas has a diameter of 512 kilometres (318 miles) and was the second asteroid to be discovered when first spotted by the German astronomer Heinrich Wilhelm Matthias Olbers on 28 March 1802. He named the object after the Greek goddess of wisdom and warfare Pallas Athena, an alternative name for the goddess Athena. Pallas has an orbital period of 1,686 days, its path around the Sun being highly eccentric and steeply inclined to the main plane of the asteroid belt, rendering it fairly inaccessible to spacecraft.

DATE	R.A.	DEC	MAG	CONSTELLATION
15 Jul 2021	23 34 15	+08 42 13	9.8	Pegasus
1 Aug 2021	23 32 45	+08 18 18	9.4	Pisces
15 Aug 2021	23 27 26	+07 00 00	9.1	Pisces
1 Sep 2021	23 16 54	+05 09 03	8.7	Pisces
15 Sep 2021	23 06 29	-01 04 34	8.5	Pisces
1 Oct 2021	22 55 24	-04 39 22	8.9	Aquarius
15 Oct 2021	22 48 29	-07 25 31	9.1	Aquarius
1 Nov 2021	22 45 20	-09 59 06	9.4	Aquarius
15 Nov 2021	22 47 27	-11 21 26	9.6	Aquarius
1 Dec 2021	22 54 42	-12 10 23	9.8	Aquarius
15 Dec 2021	23 04 42	-12 19 31	9.9	Aquarius

4 Vesta

Discovered by Heinrich Wilhelm Matthias Olbers on 29 March 1807, Vesta is one of the largest asteroids, with a diameter of 525 kilometres (326 miles) and an orbital period of 3.63 years. This is the second most massive object in the main asteroid belt, and holds the distinction of being the brightest minor planet visible from Earth. With a magnitude at best of 6 Vesta can be glimpsed with the naked eye under ideal sky conditions.

DATE	R.A.	DEC	MAG	CONSTELLATION
15 Jan 2021	11 39 50	+10 15 04	7.0	Virgo
1 Feb 2021	11 40 08	+11 43 26	6.7	Leo
15 Feb 2021	11 39 01	+13 32 22	6.3	Leo
1 Mar 2021	11 23 06	+15 34 09	6.0	Leo
15 Mar 2021	11 09 59	+17 20 40	6.1	Leo
1 Apr 2021	10 56 11	+18 34 54	6.4	Leo
15 Apr 2021	10 49 04	+18 40 41	6.7	Leo
1 May 2021	10 49 10	+17 52 16	7.0	Leo
15 May 2021	10 55 49	+16 32 08	7.2	Leo
1 Jun 2021	11 09 01	+14 19 19	7.4	Leo

20 Massalia

Massalia was discovered from the Astronomical Observatory of Naples by the Italian astronomer Annibale de Gasparis on 19 September 1852. It is the parent body of the large Massalia family which comprise of S-type (stony) asteroids located in the inner regions of the asteroid belt. To date over 6,000 Massalian asteroids are known, Massalia itself being by far the largest member of the family, with a diameter of around 145 kilometres (90 miles) and an orbital period of 3.74 years. It is interesting to note that this object was independently discovered on the following night from Marseille by the French astronomer Jean Chacornac, although full credit for the discovery was given to Gasparis. It is therefore somewhat ironic that Massalia was eventually named for the town of Marseille.

Italian astronomer Annibale de Gasparis discovered a total of nine asteroids between 12 April 1849 (10 Hygiea) and 26 April 1865 (83 Beatrix). (Wikimedia Commons)

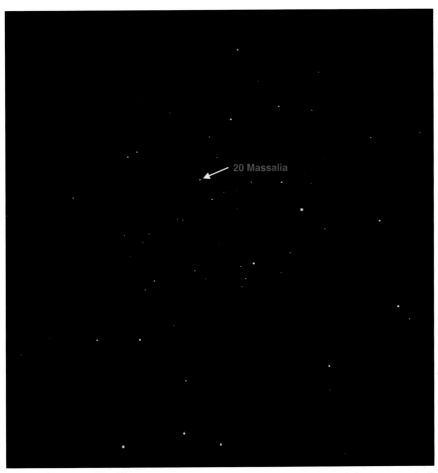

Italian astronomer Annibale de Gasparis discovered a total of nine asteroids between 12 April 1849 (10 Hygiea) and 26 April 1865 (83 Beatrix). (Wikimedia Commons)

DATE	R.A.	DEC	MAG	CONSTELLATION
1 Sep 2021	07 01 46	+22 11 19	11.1	Gemini
15 Sep 2021	07 30 54	+21 14 45	11.0	Gemini
1 Oct 2021	08 02 14	+19 50 47	10.9	Cancer
15 Oct 2021	08 27 27	+18 25 47	10.7	Cancer
1 Nov 2021	08 54 32	+16 37 06	10.5	Cancer
15 Nov 2021	09 13 10	+15 11 32	10.3	Cancer
1 Dec 2021	09 29 19	+13 49 23	10.1	Leo
15 Dec 2021	09 37 47	+13 01 24	9.8	Leo

2602 Moore

Now I thought I would include an object that is a test for those with CCD equipment, this being asteroid 2602 Moore. This is a main belt asteroid with an orbital period of 3.68 years and an orbital inclination of just over 5.54 degrees. Discovered on 24 Jan 1982 by American astronomer Edward L. G. "Ted" Bowell on photographic plates taken at the Anderson Mesa Station near Flagstaff, Arizona, this object was named in honour of the well known amateur astronomer, writer, radio and TV presenter Sir Patrick Moore, whose long-running programme The Sky at Night, first broadcast in April 1957 and still going strong today, was the main inspiration for many of us having developed an interest in astronomy.

It is perhaps fitting that this object is best placed for viewing while in Orion, poignant because this was one of Patrick Moore's favourite constellations. For those of you who have the capability to see it, asteroid 2602 Moore can be found as follows ...

DATE	R.A.	DEC	MAG	CONSTELLATION
1 Nov 2021	05 20 30	+15 01 55	16.6	Orion
15 Nov 2021	05 12 02	+14 22 07	16.1	Orion
1 Dec 2021	04 57 08	+13 43 42	16.2	Orion
15 Dec 2021	04 42 31	+13 23 24	16.2	Taurus
30 Dec 2021	04 29 10	+13 21 55	16.6	Taurus
1 Jan 2022	04 27 45	+13 23 29	16.65	Taurus

Meteor Showers in 2021

Neil Norman

Whether you are a dedicated observer or just a casual sky gazer, on any given night of the year you can reasonably expect to see several meteors dash across the sky. Quite often these will be 'sporadic' meteors, that is to say they can appear at any time and from any direction. Sporadic meteors arise when a meteoroid – perhaps a particle from an asteroid or a piece of cometary debris orbiting the Sun – enters the Earth's atmosphere and burns up harmlessly high above our heads, leaving behind the streak of light we often refer to as a "shooting star".

The meteoroids in question are usually nothing more than pieces of space debris that the Earth encounters in its path as it travels along its orbit, and range

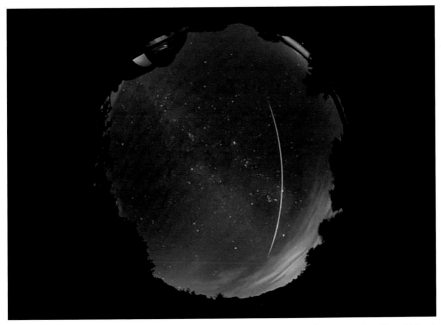

A superb sporadic meteor imaged from Nagano, Japan on 29 September 2019. (Kunihiro Shima)

in size from a few millimetres to a couple of centimetres in size. Meteoroids that are large enough to at least partially survive the passage through the atmosphere, and which reach the Earth's surface without disintegrating, are known as meteorites.

At certain times of the year the Earth encounters more organised streams of debris that produce meteors over a regular time span and which seem to emerge from the same point in the sky. These are known as meteor showers. These streams of debris follow the orbital paths of comets, and are the scattered remnants of comets that have made repeated passes through the inner solar system. The ascending and descending nodes of their orbits lie at or near the plane of the Earth's orbit around the Sun, the result of which is that at certain times of the year the Earth encounters and passes through a number of these swarms of particles.

The term 'shower' must not be taken too literally. Generally speaking, even the strongest annual showers will only produce one or two meteors a minute at best, this depending on what time of the evening or morning that you are observing. One must also take into account the lunar phase at the time, which may significantly influence the number of meteors that you see. For example, a full moon will probably wash out all but the brightest meteors.

Though impressive to the naked eye, fireballs result from pieces of debris perhaps little larger than a pea in size. Their entry speeds into the atmosphere create the friction that causes them to glow so brightly. This impressive fireball was imaged close to the Moon on 10 October 2019. (Monika Landy-Gyebnar)

The following is a table of the principle meteor showers for 2021 and includes the name of the shower; the period over which the shower is active; the ZHR (Zenith Hourly Rate); the parent object from which the meteors originate; the date of peak shower activity; and the constellation in which the radiant of the shower is located. Most of the information given is self-explanatory, but the Zenith Hourly Rate (ZHR) may need some elaborating.

The Zenith Hourly Rate is the number of meteors you may expect to see if the radiant (the point in the sky from where the meteors appear to emerge) is at the zenith (or overhead point) and if the observing conditions were perfect and included dark, clear and moonless skies with no form of light pollution whatsoever. However, the ZHR should not to be taken as gospel, and you should not expect to actually observe the quantities stated, although 'outbursts' can occur with significant activity being seen.

The observer can also make notes on the various colours of the meteors seen. This will give you an indication of their composition; for example, red is nitrogen/oxygen, yellow is iron, orange is sodium, purple is calcium and turquoise is magnesium. Also, to avoid confusion with sporadic meteors which are not related to the shower, trace the path back of the meteor and if it aligns with the radiant you can be sure you have seen a genuine member of the particular shower.

Meteor Showers In 2021

SHOWER	DATE	ZHR	PARENT	PEAK	CONSTELLATION
Quadrantids	1 Jan to 5 Jan	120	2003 EH_1 (asteroid)	3/4 Jan	Boötes
Lyrids	16 Apr to 25 Apr	18	C/1861 G1 Thatcher	22/23 Apr	Lyra
Eta Aquariids	19 Apr to 28 May	30	1P/Halley	6/7 May	Aquarius
Tau Herculids	19 May to 19 Jun	Low	73P/Schwassmann–Wachmann	9 Jun	Hercules
Delta Aquariids	12 Jul to 23 Aug	20	96P/Machholz	28/29 Jul	Aquarius
Perseids	17 Jul to 24 Aug	80	109P/Swift-Tuttle	12/13 Aug	Perseus
Draconids	6 Oct to 10 Oct	10	21P Giacobini-Zinner	7/8 Oct	Draco
Southern Taurids	10 Sep to 20 Nov	5	2P/Encke	10 Oct	Taurus
Orionids	16 Oct to 27 Oct	25	1P/Halley	21/22 Oct	Orion
Northern Taurids	10 Sep to 20 Nov	5	2004 TG_{10} (asteroid)	12 Nov	Taurus
Leonids	15 Nov to 20 Nov	Varies	55P/Tempel/Tuttle	17/18 Nov	Leo
Geminids	7 Dec to 17 Dec	75+	3200 Phaethon (asteroid)	13/14 Dec	Gemini
Ursids	17 Dec to 26 Dec	10	8P/Tuttle	21/22 Dec	Ursa Minor

Quadrantids

The Quadrantids is a significant shower that rivals the August Perseids, and peak rates can exceed 100 meteors per hour. The radiant lies a little to the east of the star Alkaid in Ursa Major and the meteors are fast moving, reaching speeds of 40 km/s. A drawback to this shower is that the period of maximum occurs over a very short time period of just 2 or 3 hours. The parent object has been identified as the near-Earth object of the Amor group of asteroids 2003 EH_1 which is likely to be an extinct comet. In 2021 a waning gibbous Moon will hinder observations, blocking out most of the fainter meteors. However, such is the activity of the shower that you should still be able to record a good number of the brighter ones.

Lyrids

These are fast moving meteors with speeds approaching 50 km/s. The rates vary, but typical values of 10 to 15 per hour are recorded. The peak falls on the night of 22/23 April with the radiant lying near the prominent star Vega in the constellation Lyra. The parent of the shower is the long-period comet C/1861 G1 Thatcher, which last came to perihelion on 3 June 1861. The period of this comet is 415 years and it will next approach perihelion in 2280. The glare from a waxing gibbous Moon will tend to hinder observations this year by blocking out all but the brightest meteors from this shower.

Eta Aquariids

This is one of the two showers associated with 1P/Halley and is active for a full month between 19 April and 21 May with meteors attaining speeds of around 65 km/s. The radiant lies just to the east of Alpha (α) Aquarii (Sadalmelik) and up to 30 meteors per hour are normally expected during the period of maximum activity, which occurs on the night of 6/7 May. This year a waning crescent Moon should present meteor watchers with ideal observing conditions.

Tau Herculids

Appearing to originate from the star Tau (τ) Herculis, this shower runs from 19 May to 19 June with peak activity taking place on 9 June. The Tau Herculids were first recorded in May 1930 by observers at the Kwasan Observatory in Kyoto, Japan. The parent body is the comet 73P/Schwassmann–Wachmann, discovered on 2 May 1930 by the German astronomers Arnold Schwassmann and Arno Arthur Wachmann during a photographic search for minor planets being carried out from Hamburg Observatory in Germany.

73P/Schwassmann–Wachmann has an orbital period of 5.36 years. During 1995 the comet began to fragment, and by the time of its 2006 return, at least eight individual fragments were observed (although the Hubble Space Telescope spotted dozens more). 73P/Schwassmann–Wachmann appears to be close to total disintegration. Although the observed rates of meteors from this shower are low, peak activity (such as it is) in 2021 occurs at around the time of new Moon, thereby presenting dedicated meteor-watchers with an opportunity to record members of this elusive shower.

Delta Aquariids

Linked to the short-period sungrazing comet 96P/Machholz, the Delta Aquariids coincide with the much more prominent Perseids, although Delta Aquariid meteors are generally much dimmer than those associated with the Perseids, making identification much easier for the observer.

The radiant lies to the south of the Square of Pegasus, close to the star Skat in Aquarius and a little to the north of the bright star Fomalhaut in Piscis Austrinus. Although never very high above the horizon as seen from mid-northern latitudes, the radiant is well placed for those observers situated in the southern hemisphere. The shower peaks during the early morning of 29 July, and in 2021 a waning gibbous Moon will be a problem as its glare will tend to drown out many of the faintest meteors.

Perseids

The Perseids are fast moving, clocking in at speeds approaching 60 km/s, due to the parent comet (109P/Swift-Tuttle) being in a retrograde orbit. This is a potentially gorgeous spectacle, with meteors appearing as soon as night falls and with up to 80 meteors per hour at their peak, which occurs on the night of 12/13 August. Large fireballs are often observed, with some even seen to cast shadows. A note of caution though - this is a busy time of year for showers, so inexperienced observers should ensure they follow the trajectories of any meteors seen back to the radiant point in the northern reaches of Perseus. The waxing crescent Moon will set early in the evening, leaving dark skies for what should be an excellent show.

Two Perseid meteors captured on the morning of 7 August 2019 (Monika Landy-Gyebnar)

Southern Taurids/Northern Taurids

This shower is associated with the periodic comet 2P/Encke, with both southern and northern components, the southern hemisphere encountering it first followed by the northern hemisphere later on. The expansive size of this stream is due to a much larger object having disintegrated in the past, leaving a large fragment (2P/Encke) as the main survivor and the remaining debris spread out along the orbit. Larger pieces are believed to lurk within, and many fireballs are seen with this shower. The northern Taurid radiant lies near the Pleiades (M45) open star cluster. Although there are low ZHRs for both components (between 5 and 10 meteors per hour), they can be beautiful to watch as they appear to glide effortlessly across the sky at speeds of 28 km/s. It is unusual in that it consists of two separate streams, the first produced by from debris left behind by comet 2P/Encke and the second emanating from dust grains left behind by asteroid 2004 TG_{10} which may be a fragment of 2P/Encke. The shower runs annually from 10 September to 10 December, the southern stream peaking this year on the night of 10 October and the northern on 12 November.

Draconids

Also known as the Giacobinids, the Draconid meteor shower emanates from the debris left behind by periodic comet 21P/Giacobini-Zinner. The duration is from 6 October to 10 October, the shower peaking on the night of 7/8 October. The ZHR of this shower can vary. For example, although the hourly rates for 1915 and 1926 appear to have been relatively low, significant displays were seen in 1933 (with hourly rates of several hundred) and 1946 (with thousands of meteors per hour being recorded). Increased activity was also noted in 1998, 2005 and 2012. Radiating from a point near the "head" of Draco, the meteors from this shower travel at a relatively modest 20 km/s. They are generally quite faint, although this year the Moon will be less than two days old, producing dark skies for what may be a moderately good display.

Orionids

This is the second of the meteor showers associated with 1P/Halley, occurring between 16 and 27 October with a peak on the night of 21/22 October. The radiant of this shower is situated a little way to the north of the conspicuous red super giant star Betelgeuse in the shoulder of Orion the Hunter, and the Orionids are best viewed in the early hours when the constellation is well placed. The velocity of the meteors entering the atmosphere is a speedy 67 km/s. During the time of peak activity the Moon will be just past full phase, and its glare will be a problem for those wishing to catch a glimpse of this shower this year.

Leonids

This is a fast moving shower with atmosphere impact speeds of 72 km/s and with particles varying greatly in size, including a lot of larger sized pieces of debris with diameters in the order of 10mm and masses of around half a gram. These can create lovely bright meteors that occasionally attain magnitude −1.5 or better. It is interesting to note that each year around 15 tonnes of material is deposited over our planet from the Leonid stream.

The parent of the Leonid shower is the periodic comet 55P/Tempel Tuttle which orbits the Sun every 33 years and which was last at perihelion in 1998 and is due to return in late-May 2031. The radiant is located a few degrees to the north of the bright star Regulus in Leo. The Zenith Hourly Rates vary due to the Earth encountering material from different perihelion passages of the parent comet. For example, the storm of 1833 was due to the 1800 passage, the 1733 passage was responsible for the 1866 storm and the 1966 storm resulted from the 1899 passage (for additional information see Courtney Seligman's article *Cometary Comedy and Chaos* in the *Yearbook of Astronomy 2020*). At the time of maximum activity, the night sky will be subject to the glare of a nearly full Moon, resulting in the light from all but the brightest meteors being drowned out this year.

Geminids

The Geminids were first recorded as recently as 1862 which indicates that the stream has only recently been perturbed near to the Earth by Jupiter. The parent of the shower is the object 3200 Phaethon, an asteroid that is in many ways behaving like a comet. Discovered in October 1983, this rocky 5-kilometre wide object is classed as an Apollo asteroid and has an unusual orbit that takes it closer to the Sun than any other named asteroid. Classified as a potentially hazardous asteroid (PHA), 3200 Phaethon made a relative close approach to Earth on

A Geminid meteor captured over Newcastle on 11 December 2015. (David Blanchflower)

10 December 2017, when it came to within 0.069 AU (10.3 million kilometres / 6.4 million miles) of our planet.

Geminid meteors travel at relatively modest speeds of 35 km/s and disintegrate at heights of around 40 kilometres above the Earth's surface. The shower radiates from a point close to the bright star Castor in Gemini and peaks on the night of 13/14 December. This is considered by many to be the best shower of the year, and it is interesting to note that the number of observed meteors appears to be increasing each year. As far as the 2021 shower is concerned, a waxing gibbous Moon will block out most of the fainter meteors although the Geminids should still put on a moderately good show for the keen meteor observer.

Ursids

Discovered by William F. Denning during the early twentieth century, this shower is associated with the Jupiter family comet, 8P/Tuttle. It has been noted that outbursts occur when the comet is at aphelion as some meteoroids get trapped in a 7/6 orbital resonance with Jupiter. The shower radiant lies near Beta (β) Ursae Minoris (Kochab). With relatively slow speeds of 33 km/s, the Ursids will appear to move gracefully across the sky. The Ursids peak on the night of 22 December when the Moon will have just passed full phase, blocking out the light from many of the fainter meteors.

Article Section

Astronomy in 2020

Rod Hine

Planck Legacy Data – a Twist in the Tail?

The Planck Observatory, which operated from 2009 to 2013, produced the definitive data on the Cosmic Microwave Background radiation (CMB) which is one of the fundamental pillars of the "Standard Model of Cosmology". Widely accepted, the model explains how the universe was created from pure energy some 13.8 billion years ago and is now formed from a mixture of dark energy, cold dark matter and ordinary matter in very specific proportions. To explain the detailed path from the instant of creation to the currently observed distribution of stars and galaxies there are many assumptions and theories to make the model work satisfactorily, few more important than the "period of inflation".

It was American theoretical physicist and cosmologist Alan Harvey Guth, then working at Cornell University, USA, who first shared his ideas at a seminar in 1980 to show how a very short but intense episode of exponential expansion of the universe could perhaps explain some puzzling features of the CMB. Guth and others continued to develop the theory further and were able to show that it accounted for the isotropic nature of the universe; it showed why the CMB was distributed so evenly across the sky and why the universe appears to be flat. In this context, flatness means that there is no curvature and presumably the universe is infinite in extent. For a prosaic demonstration that we live in a flat part of the universe we can merely observe that the interior angles of a triangle add up to 180° – as simple as that.

Given that a flat universe is so much a part of the Standard Model, some recent work by Alessandro Melchiorri at the Sapienza University of Rome has caused some consternation. Melchiorri and his colleagues studied the recently released Planck Legacy 2018 final data set for signs of gravitational lensing. This phenomenon arises where light rays are bent and focused by passing close to massive objects located between the source and the observer, and in the CMB data gives a measure of the amount of dark matter. Their paper, published in

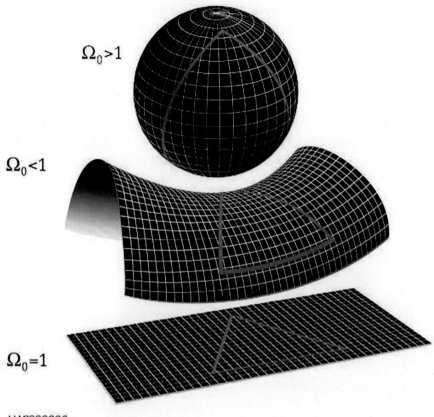

$\Omega_0 > 1$

$\Omega_0 < 1$

$\Omega_0 = 1$

MAP990006

The local geometry of the universe is determined by whether the density parameter Ω is greater than, less than, or equal to 1. From top to bottom: a spherical universe with $\Omega > 1$; a hyperbolic universe with $\Omega < 1$; and a flat universe with $\Omega = 1$. These depictions of two-dimensional surfaces are merely easily-visualizable analogs to the three-dimensional structure of (local) space. (NASA Official: Gary Hinshaw)

Nature Astronomy on 4 November 2019, shows that there could be a lot more dark matter than hitherto assumed, and that this extra dark matter could pull the universe into a finite sphere. They assign a confidence level of more than 99% that the results favour a closed universe with positive curvature.

If this is correct, and the universe is not flat after all, then numerous other problems arise for cosmologists. It has been difficult enough in a flat universe to account for the increasing rate of expansion, attributable to dark energy, and will be even tougher for a closed spherical universe. For many cosmologists, the hope is that this observation is some kind of statistical fluke and can be rebutted.

There are no other observations to back it up so far, but it may not be the end of the story when the Simons Observatory in Chile begins operations in the early 2020s. Funded by a generous grant of $40 million from the Simons Foundation of New York, USA, together with contributions from many participating universities, the site in Cerro Toco, Atacama Desert, Chile, will host a set of radio telescopes to measure the CMB with utmost precision at frequencies from 27 GHz to 280 GHz. The CMB peaks at 160 GHz so the new telescopes will be able to cover the most interesting parts of the spectrum but with much finer resolution than the Planck satellite. In addition to the main

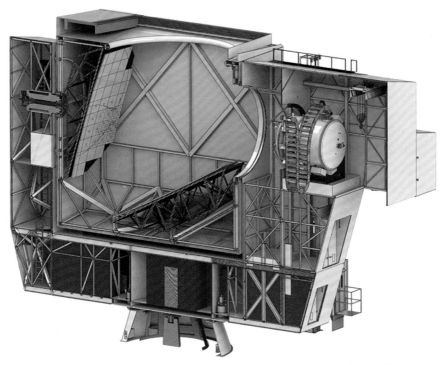

A cross section through the Simons Observatory Large Aperture Telescope showing the mirrors housed in the elevation structure. The white cylinder on the right is the 2.4 meter diameter cryostat. (Wikimedia Commons / Sdicker – Own Work)

Large Aperture Telescope (LAT) with a 6 metre mirror there will be three Small Aperture Telescopes (SAT) of 0.5 metre aperture to form an array. Overall the resolution will be at least an order of magnitude better than Planck with its under 2 metre primary mirror. The instruments in the LAT will be housed in a huge cryostat to reduce noise and increase sensitivity. The unusual design, the so-called "crossed Dragone" configuration, allows a compact fully-steerable mirror to deliver the signals to the horizontally mounted cryostat fixed on the structure and therefore more accessible and protected than conventional mounting at the prime focus of a paraboloid.

Gravitational lensing of the CMB is high on the observing priority for the Simons Observatory and it will be fascinating to see whether Melchiorri's results hold good or can be dismissed. There are several other anomalies in the Standard Model, for example the unexplained variation in the Hubble Constant when calculated from different metrics. Another major spanner in the works could undermine the Standard Model and cause a genuine cosmological crisis.

The Mystery of Variable Quasars

Since the early 1960s the Santa Catalina Mountains close to Tucson, Arizona, USA, have been home to an observatory run by the University of Arizona. Largely inspired by the famous Gerard Kuiper of that same University, the observatory was moved from its original site on Mount Bigelow to Mount Lemmon in 1971. At that time there were two telescopes on Mount Lemmon, one of 1.5 metre aperture and another of 1 metre. In 1998 a team came together for a new project called the Catalina Sky Survey (CSS). Led by staff scientist Steve Larson and two undergraduates, Tim Spahr and Carl Hergenrother, they managed also to secure access to the unused 0.7 metre Schmidt Telescope that remained on Mount Bigelow and began an extensive photographic survey of objects at higher ecliptic latitude. With help from NASA they began to search for NEOs (Near Earth Objects) and eventually established the Siding Spring Survey in New South Wales, Australia, also dedicated to NEO studies.

In addition to discovering over three hundred comets and numerous asteroids, the same telescopes provide data for the Catalina Real-time Transient Survey (CRTS), founded in 2007 and operated from California Institute of Technology (Caltech), which notifies observers immediately of any optical transient events detected in the images.

This artist's impression of one of the most distant, oldest, brightest quasars ever seen is hidden behind dust. The quasar dates back to less than one billion years after the big bang. The dust is also hiding the view of the underlying galaxy of stars that the quasar is presumably embedded in. (NASA/ESA/G.Bacon, STScI)

A recent publication by a group from Caltech led by Matthew Graham has used the CSS and CRTS data to try to resolve one of the strangest features of quasars. Conventional wisdom describes quasars as supermassive black holes at the core of incredibly distant galaxies, devouring gas and dust from its surroundings and shining as a beacon across billions of light-years. Such a scenario should take tens of thousands of years to deplete the local supply of material yet some quasars have been observed to fade out in timescales of as little as a year, leaving an unexceptional ordinary-looking galaxy in its place. Some other quasars have suddenly flared into existence from otherwise normal galaxies, again with suspiciously short timescales.

A group at JPL, Pasadena, led by Chelsea MacLeod of Harvard Smithsonian Centre for Astrophysics, Cambridge, Massachusetts, had already compiled a list of more than 200 highly variable quasars and active galactic nucleus (AGN) objects that could be described as "changing-look quasars" (thus adding CLQ to the list of astronomical acronyms). Published in early 2019 in the *Astrophysical Journal*, they speculated on the many possible explanations and specifically on whether the variations were intrinsic to the quasar or extrinsic, i.e. arising from the quasar's surroundings.

Graham and his colleagues systematically searched the CSS and CRTS data for quasars which exhibited different variability behaviours such as periodic variability, major flaring episodes and extreme variations in spectral line properties. By studying such objects in all possible wavelengths from infrared to visible, they have observed that when the visible light from a quasar changes the same change is echoed afterwards in the infrared. Since the visible light mostly emanates from the accretion disk and the infrared comes from the more distant surrounding clouds of cooler dust this strongly suggests that the visible light warms the dust which then re-radiates the energy in the infrared, and so the rapid changes are due to changes in the accretion disk itself, as unlikely as that seems. Previous explanations had suggested that quasars might have been eclipsed by dust or magnified by chance gravitational lensing events but, for these quasars at least, that now seems unlikely. The extreme variations are indeed intrinsic to the quasar.

Graham has gone so far as to classify at least 73 quasars as "changing-state quasars" (CSQ) from hundreds of candidate CLQs and has shown that this model can account for the rapidly-varying brightness by heating and cooling wave-fronts propagating through the accretion disk in the observed timescales. It still isn't clear exactly what causes the rapid variations in the first place, and there may be several different causes according to John Ruan of McGill University, Montreal, Canada. Based on the wide diversity of the examples he says, "I would not be surprised at all if changing-look quasars are due to a variety of things." Ruan was not involved in the study and is a post-doctoral Fellow at McGill specialising in AGN variability and accretion disk transitions.

First Results from NASA's Parker Solar Probe

Launched in August 2018, the Parker Solar Probe, named for Emeritus Professor Eugene Parker of the University of Chicago, has sent back results of its first hazardous flyby of the Sun. In 24 successive approaches over the next seven years or so, the spacecraft will fly ever closer, using Venus's gravity to dive through the Sun's intense heat and radiation to within 3.8 million miles (about 6 million kilometres) of the photosphere. At its closest approach it will be travelling at 430,000 mph (700,000 km/h) and the heat shield will reach 1,377°C. Key to the spacecraft's survival is the 12 centimetre thick heat shield which will always be facing the Sun and will keep the rest of the craft at sensible room temperature.

This almost suicidal mission will make measurements of magnetic fields, radiation, the corona and the solar wind to try to discover exactly how the outer atmosphere of the Sun works. Scientists have pondered long and hard to explain how the corona is heated to around a million degrees when the temperature of the Sun's visible surface is a balmy 5780 Kelvin.

Early results released by NASA show that the solar wind close to the Sun is far more dynamic and turbulent than expected and has surprised the scientists.

SDO/AIA 193Å 20181027

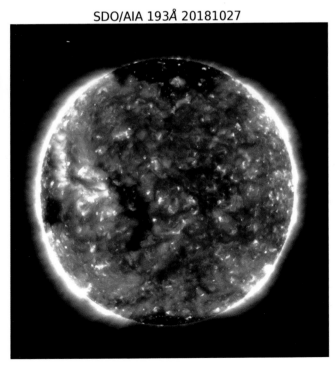

NASA's Parker Solar Probe observed a slow solar wind flowing out from the small coronal hole, seen here as the long, thin black spot on the left side of the Sun in this image captured by NASA's Solar Dynamics Observatory on 27 October 2018. While scientists have long known that fast solar wind streams flow from coronal holes near the poles, they have not yet conclusively identified the source of the Sun's slow solar wind. (NASA/SDO)

"The complexity was mind-blowing when we first started looking at the data," said Stuart Bale, the University of California, Berkeley, lead for Parker Solar Probe's FIELDS instrument suite, which studies the scale and shape of electric and magnetic fields. "Now, I've gotten used to it. But when I show colleagues for the first time, they're just blown away." From Parker's vantage point 15 million miles from the Sun, Bale explained, the solar wind is much more impulsive and unstable than what we see near Earth.

It will take a long time to analyse the complex interactions observed between the plasma and the electric and magnetic fields around the spacecraft. There is evidence of rapid reversals in the magnetic field lasting from a few seconds to several minutes and sudden fast-moving jets of plasma swirling around. Understanding these turbulent flows will be vital to discover how energy and particles propagate away from the Sun.

Other curious features measured for the first time are the rotation of the solar wind closer to the Sun and the apparent lack of dust very close to the Sun.

"This dust-free zone was predicted decades ago, but has never been seen before," said Russ Howard, principal investigator for the WISPR suite – short for Wide-field Imager for Solar Probe – at the Naval Research Laboratory in Washington, D.C. "We are now seeing what's happening to the dust near the Sun."

It was in 1958 that solar astrophysicist Eugene Parker first theorized about the existence of the solar wind, and as recently as 2018 was awarded the American Physical Society Medal for Exceptional Achievement in Research. He is the only living person to have a spacecraft named after him, the mission name having been changed by NASA from "Solar Probe Plus" in 2017.

Picture of Black Hole gets prestigious Physics World Award

Astronomers working on the Event Horizon Telescope project were honoured by the award of the "Physics World 2019 Breakthrough of the Year". After decades of theorizing about the very existence of black holes the moment finally arrived when we could actually see what a black hole looks like. We can't

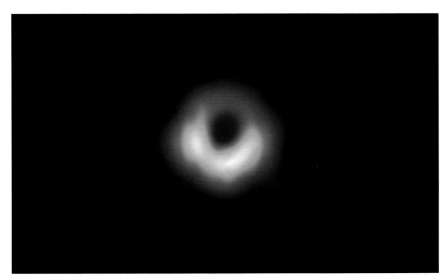

Using the Event Horizon Telescope, scientists obtained an image of the black hole at the centre of galaxy M87, outlined by emission from hot gas swirling around it under the influence of strong gravity near its event horizon. (Event Horizon Telescope collaboration et al)

see the actual black hole itself but the image of the crescent-shaped ring of hot gas was clearly just as Einstein's general relativity predicted. The image was derived from data collected by eight radio telescopes at six different locations – a formidable task in data processing alone, to say nothing of the precision required to precisely synchronise the observations of M87 some 55 million light years distant from Earth.

Solar System Exploration in 2020

Peter Rea

This article was written in the autumn/winter of 2019. As all the missions mentioned are active, the status of some missions may change after the print deadline. There has been much activity since the last Yearbook and much of this article takes place in 2019 but continues into 2020.

In the beginning, Israel went to the Moon

Israel was set to become the fourth nation (after Russia, the USA and China) to land a spacecraft on Moon. Back in 2007, Google offered the Lunar X Prize, worth at the time $20 million, to any private organization who could land a spacecraft on the Moon then travel a minimum of 500 metres and return images and video. The deadline was 2017 but later extended to 2018. The prize was withdrawn in 2018 when no company achieved a landing. However, the Israeli company SpaceIL (established in 2011) continued with their spacecraft and booked a ride share flight on a SpaceX Falcon 9 rocket which was launched on 22 February 2019 (GMT). The spacecraft, now named Beresheet (Hebrew

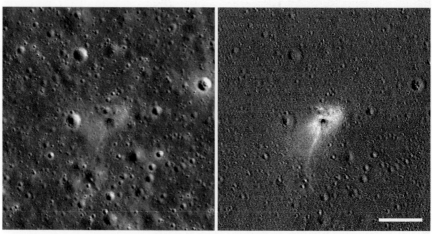

Beresheet crash site imaged from lunar orbit. (NASA/GSFC/Arizona State University)

"Genesis"), went into orbit with a communications satellite destined for a geostationary transfer orbit. From this orbit Beresheet performed a series of orbit raising manoeuvres to gradually increase the orbital apogee to a point where lunar gravity took over and pulled it toward the Moon. It arrived in lunar orbit on 4 April 2019 and prepared for a landing on the Mare Serenitatis. The deorbit burn occurred on 11 April but control issues on the way down caused Beresheet to crash on the Moon. Further details of the mission can be found at **www.spaceil.com**

Sample collecting on Ryugu

Launched on 3 December 2014 the Japanese Hayabusa 2 mission arrived at the near-Earth asteroid 162173 Ryugu on 27 June 2018. Ryugu is a carbonaceous asteroid and is expected to have been formed during the early history of the solar system. Collecting some pristine samples of Ryugu is a major goal of the mission. Studying these samples is expected to provide knowledge on the origin and evolution of the inner planets. On 21 September 2018 two small rovers were ejected from Hayabusa 2 to free fall to the surface of Ryugu. A German supplied rover called MASCOT was deployed to the surface on 3 October

Hayabusa 2 second touchdown point on Ryugu. (JAXA, University of Tokyo, Kochi University, Rikkyo University, Nagoya University, Chiba Institute of Technology, Meiji University, University of Aizu and AIST)

2018. The first surface sample retrieval took place on 21 February 2019 and on 5 April Hayabusa 2 released an impactor into the surface of Ryugu which created a small crater allowing instruments on the orbiter to study below the surface. A second sample collection occurred on 11 July 2019. After completing all science objectives at the asteroid, Hayabusa 2 fired up the Ion thrusters on 13 November 2019 (GMT) to slowly accelerate the spacecraft away from Ryugu and start the return journey to Earth. The sample return capsule will parachute down to a soft landing in Woomera, Australia in December 2020. The mission website can be found at **www.hayabusa2.jaxa.jp/en**

Sample collecting on Bennu

The OSIRIS-REx (Origins, Spectral Interpretation, Resource Identification, Security, Regolith Explorer) mission from the USA has been in orbit around the asteroid 101955 Bennu since late December 2018 after a journey of more than two years. The initial orbit around Bennu took the spacecraft down to around 2 kilometres above the surface. It was from here over the next few months a complete survey of the asteroid was performed. The surface of Bennu is almost totally cluttered with loose rubble and finding a suitable landing site

Close-up of Bennu. Finding a clear safe area to collect a sample is not easy! (NASA, Goddard Space Flight Center, University of Arizona)

to collect samples is proving a challenge. By altering the inclination of the orbit, observations of all latitudes were conducted. By August 2019 the list of potential sampling sites had been narrowed down to four. Once the final site is chosen, OSIRIS-Rex will be manoeuvred into a position above the landing site from where it will slowly move down to the surface in much the same way as Hayabusa 2 has done at Ryugu. A special instrument known as TAGSAM or Touch-And-Go Sample Acquisition Mechanism is extended below the spacecraft and will be the only part to touch Bennu. As soon as the tip of TAGSAM touches the surface a burst of nitrogen gas will blow part of the surface material into the sampler head. This can be repeated up to three times. Scientists are looking to collect at least 60 grams of surface material which will be placed into a sample return capsule. OSIRIS-Rex is due to leave Bennu in March 2021 for a return trip to Earth, arriving in September 2023 with a parachute landing at the Utah Test and Training Range in the USA. The mission website can be found at **www.asteroidmission.org**

Moles on Mars and other Insights

The NASA InSight mission to Mars landed successfully on 26 November 2018. A stationary lander unlike recent rovers it carried two principal experiments designed to give an insight (pun intended) into the interior structure of Mars. A seismometer to listen for marsquakes was deployed onto the surface of Mars on 19 December 2018 and was fully commissioned by early February 2019. During the summer of 2019 the first data indicating possible marsquakes was received. This is an ongoing investigation and hopefully by the time this publication goes on sale more will have been discovered.

The other principal experiment was the Heat Flow and Physical Properties Package. This is essentially a self-drilling probe called a "mole" designed to drill down to a depth of perhaps five metres. The package was deployed onto the surface on 12 February 2019 and the mole started pushing down through the surface on 28 February 2019. After reaching no more than 30 centimetres the mole unexpectedly stalled. Over the next few months a recovery plan was worked on and numerous tests on a spare unit at the Jet Propulsion Laboratory in California were carried out. As this recovery plan is ongoing at the time of writing, interested readers should consult the mission website for the very latest situation which can be found at **mars.nasa.gov/insight**

InSight mole imaged
on surface by camera
on robotic arm.
(NASA/JPL-Caltech)

Information Interlude

From time to time the various mission controls loose contact with their respective spacecraft for perfectly natural reasons. This occurs when the Earth and the planet involved are on opposite sides of the sun. The signal from or to the spacecraft is lost in the intense radiation surrounding the sun, so no communication is possible for a week or two. This planetary alignment is known as solar conjunction and is shown in the accompanying diagram.

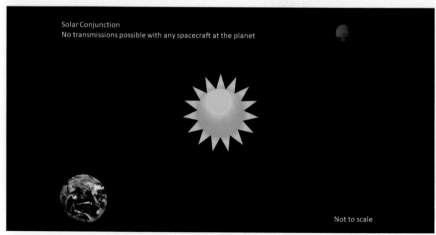

Solar conjunction. (Peter Rea)

Six wheels to Jezero Crater

The next Mars lander builds on the success of the Mars Science Laboratory named Curiosity, which landed in Gale crater on Mars on 6 August 2012, and as at time of writing is still working well. It benefits from using the same tried and tested design as Curiosity and viewed side by side Curiosity and the Mars 2020 rover look almost identical. The science instrument package on Curiosity was designed for geological studies. However, the instruments on the Mars 2020 rover, though capable of geological studies, are more suited to the study of possible past life on Mars. The four mission goals are to: determine whether life ever arose on Mars; study the climate of Mars; characterise the geology of Mars; and prepare for human exploration. The target is the 49 kilometre wide Jezero crater situated on the boundary between the dark albedo feature Syrtis Major Planitia and Isidis Planitia, the resting place for the UK-built Beagle 2 lander. Jezero was chosen because there is evidence from orbital images that the

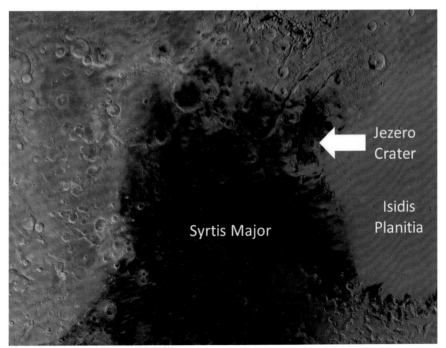

Location of Jezero crater, north east of Syrtis Major. (Background image Wikimedia Commons adapted by Peter Rea)

crater was once filled with water. A fan delta has been observed that is rich in clays which form in the presence of water. Did this area once support ancient microbial life?

The rover will also carry as a technology demonstrator a 1.8 kg drone helicopter known as the Mars Helicopter Scout. Fitted underneath the rover for landing, it will be released from the rover and placed on the surface of Mars between the wheels. The rover will then drive slowly forward to reveal the solar and battery powered drone.

Mars 2020 rover and Mars Helicopter Scout. (NASA/JPL-Caltech)

In anticipation of a manned mission to Mars sometime in the future the 2020 rover will carry, among many other experiments, an instrument capable of producing small quantities of oxygen from the tenuous Martian atmosphere. This is only a small-scale experiment, but if successful could prove very important if it can be scaled up for future manned missions.

It will also collect surface samples and leave them on the surface in special containers for a future sample return mission to collect them and return the containers back to Earth.

This rover will be packed with cameras, far more than on any other mission to Mars. The Mars Science Laboratory Curiosity had a total of 17 cameras yielding some magnificent views, not just of Mars but of the landing back in 2012. The Mars 2020 rover will carry 23 cameras. The views of landing and parachute deployment should be breath taking. Launch is targeted for July 2020 with a landing on 18 February 2021. The mission website, giving a comprehensive description of all the science instruments, can be found at **mars.nasa.gov/mars2020**

Rosalind Franklin goes to Mars

As part of their ongoing ExoMars programme, the European Space Agency (ESA) in conjunction with Roscosmos in Russia was due to send a lander and rover to Mars in July 2020, but in March 2020 a decision was made to delay the mission to the next launch opportunity in 2022. The Rover is named 'Rosalind Franklin' after the English chemist and x-ray crystallographer whose work would assist in the discovery of the double helix structure of deoxyribonucleic acid (DNA). ESA has not yet successfully landed on Mars and readers may recall that back in 2016 the ESA Trace Gas Orbiter mission carried an experimental lander called Schiaparelli. The lander entered the Martian atmosphere on 16 October 2016. The initial entry went according to plan and a parachute was successfully deployed. However shortly after this, issues with the computer caused an early release of the parachute and the retro propulsion system thrusters only fired for a few seconds before being commanded off. This caused the lander to impact the surface at 540 kph. However much was learnt from this landing attempt and the reason for the failure is understood giving confidence for the 2022 lander.

Rosalind Franklin rover on Mars showing landing stage in middle ground. (ESA/ATG medialab)

The method of landing uses the same technique as the Schiaparelli lander – heat shield, parachute and powered descent. Roscosmos will provide the lander which carries a significant suite of science instruments for in situ measurements. Data from these instruments will be relayed back to Earth via ESA's Trace Gas Orbiter. The rover is principally built in the UK at Airbus Defence and Space in Stevenage. Landing is targeted for the Oxia Planum region which lies east of Ares Vallis and Chryse Planitia, where Viking 1 landed back in 1976. Once the rover is deployed onto the surface of Mars for independent studies the lander will pursue a stationary investigation of the landing site with its own suite of instruments. I think Rosalind Franklin would have been proud.

The mission website giving a comprehensive description of all the science instruments can be found at **exploration.esa.int/mars/48088-mission-overview**

The Long and Winding Road (to Mercury)

The European Space Agency's BepiColombo mission to Mercury is well on its way to an encounter with the innermost planet in the solar system, Mercury. Launched on 20 October 2018 it will spend the first part of its complicated journey to Mercury in the vicinity of Earth's orbit. The flight plan calls for BepiColombo to re-encounter Earth on 6 April 2020 followed by a flyby of Venus on 12 October of the same year. A second flyby of Venus on 11 August 2021 will set up the first of six Mercury flybys on 2 October 2021. The following five Mercury flybys will circularise the orbit in preparation for Mercury orbit

The BepiColombo Mercury Transfer Module (foreground, with two large solar wings) carries the Mercury Planetary Orbiter (middle, with one solar wing pointing up) and the Mercury Magnetospheric Orbiter (hidden inside the solar shield, on the far side) to Mercury. The onboard ion propulsion, preferred over chemical propulsion for its greater efficiency, can be seen in this image at the rear of the Mercury Transfer Module. (ESA/ATG medialab)

insertion on 5 December 2025. The mission can be followed at **sci.esa.int/bepicolombo**

India Returns to the Moon for a Landing Attempt

On 6 September 2019 the Indian Space Research Organisation attempted a landing close to the south pole on the moon. The Chandrayaan-2 spacecraft had been launched on 22 July 2019 and initially entered an elliptical orbit around the Earth. Successive firings of the onboard propulsion system through to 6 August gradually increased the apogee to almost 143,000 kilometres. On 13 August a further firing of the propulsion system pushed Chandrayaan 2 out of Earth orbit and onto a trajectory toward the Moon. On 20 August the spacecraft arrived at the moon in a highly eccentric orbit which over the next few days was reduced to a more circular one. By 1 September the orbit had a perilune (the point at which a spacecraft in lunar orbit is closest to the moon) of 127 kilometres. On 3 September the perilune was as low as 35 kilometres. On 6 September the Vikram lander separated from the orbiter or mother ship and prepared for the retro fire which commenced the landing sequence. This writer used the World Wide Web to connect to the Indian mission control centre to follow the landing sequence live. Landing was scheduled at 9.23 pm UK time. On the screens in front of the controllers was a plot which showed the path of the lander as it approached the landing site. All went well until the lander passed 2 kilometres from the surface. I could see for myself a deviation from the nominal approach and the path went more vertical. This did not look right. As the time of nominal landing passed, I could see one controller with his head in his hands. It was obvious the landing had not occurred as planned, although it was still a great effort by the Indians who realised, just like the Israelis did with their Beresheet lunar lander, that the difference between success and failure can be agonisingly small. All is not lost as the orbiter with a suite of instrument remains in orbit to conduct science investigations of the Moon over the next few years. It may be a setback for the Indians, but they will be back at the Moon soon to try and court favour with the science community. The Chandrayaan-2 mission website can be found at **www.isro.gov.in/chandrayaan2-home-0**

Something To Look Forward To ...

NASA has just been given the go ahead for a new mission to Jupiter. The primary objective of the Europa Clipper mission is to study Europa, which could have an internal ocean below the icy surface. The mission could launch as early as 2023.

The European Space Agency is also going to Jupiter with their Jupiter Icy Moons Explorer (JUICE) mission. Due for launch in 2022 it will fly past Europa twice and Callisto many times before going into orbit around Ganymede.

Dragonfly is a mission to send a robotic drone to Titan, the largest moon of Saturn. The rotorcraft will fly in short hops to various locations on Titan. It is due for launch in 2026 for a landing on Titan in 2034.

The American space agency NASA and the European Space Agency ESA are looking to collaborate on a Mars sample return mission. The mission would look to collect samples left on Mars by the Mars 2020 rover, and could launch as early as 2026. See article above.

As always, Solar System exploration continues to excite and inspire, and next year promises to be no different.

Anniversaries in 2021

Neil Haggath

Johannes Kepler (1571–1630)

This year sees the 450th anniversary of the birth of Johannes Kepler, who has been described as the world's first astrophysicist. While his contemporary Galileo established the methods of what we now call experimental physics, Kepler was the first to realise that the motions of astronomical bodies could be described by the laws of physics and mathematics.

He was born on 27 December 1571 in Weil der Stadt in Württemberg, now part of Germany. His upbringing was difficult; his father left when he was young, and his mother was tried for witchcraft, though Johannes managed to get her acquitted. At the age of five, his mother took him outdoors to see a bright comet, and he was fascinated by astronomy from then on.

Portrait of Johannes Kepler by an unknown artist, from a lost original of 1610 in the Benedictine monastery in Kremsmünster. (Wikimedia Commons)

While studying at the University of Tübingen, Kepler became an early supporter of Copernicanism – though as Copernicus himself had acknowledged, there were discrepancies between the observed motions of the planets and those predicted by his theory. He wrote a book proposing a model of the structure of the Solar System, involving the regular solids of classical Greek geometry, which was frankly silly, and had more to do with ancient Greek mysticism than with science. But he soon realised that it didn't work, and in the manner of a true scientist, abandoned it and started again.

In 1600, Kepler was invited to Prague to work as assistant to Tycho Brahe (1546–1601). Tycho was perhaps the most meticulous observer of the pre-telescopic era, but was no theoretician, and refused to accept Copernicanism. Kepler, in contrast, was not much of an observer, but a brilliant mathematician and theoretician. After Tycho's death in 1601, Kepler succeeded him as Imperial Mathematician, and inherited all his observations of the planets – the most accurate to date – and set about trying to reconcile them with Copernican theory. He eventually realised that the discrepancies could be explained by assuming that the planets' orbits were not circular, as Copernicus had assumed, but elliptical.

Between 1609 and 1618, Kepler published his famous Three Laws of Planetary Motion. His Third Law, which states that the square of a planet's orbital period is proportional to the cube of its mean orbital radius, was a forerunner of the inverse square law of gravity, and proved the key to determining the size of the Solar System; given that relationship, it was only necessary to measure the Earth's distance from the Sun, and those of all the other planets could be calculated.

Kepler was forced to leave Prague after his patron, Emperor Rudolph, was deposed. He went to Austria, and later to Silesia, now part of Poland, where he died on 15 November 1630.

Remarkably, Kepler even wrote what was possibly the very first work of what we would now call science fiction. His last book, published posthumously, was a novel entitled the *Somnium*, or Dream, about a journey to the Moon!

Giovanni Riccioli (1598–1671)

When astronomers first began mapping the moon in the 17th century, several different naming systems were introduced for its surface features. But the system which survives to this day was the work of a man who died 350 years ago, the Italian Jesuit astronomer Father Giovanni Battista Riccioli, SJ.

Riccioli was born on 17 April 1598 in Ferrara, Italy. He entered the Society of Jesus at age 16, and was ordained as a priest in 1628. In 1636, he was sent to Bologna to become Professor of Theology – but he maintained a lifelong passion for physics, mathematics and astronomy, and built an observatory in Bologna.

He performed extensive studies of pendulums and falling bodies. The latter, like that of Galileo before him, involved dropping objects from a tall building – in this case, the 97-metre Torre de Asinelli in Bologna. He made the first reasonably accurate measurement of the acceleration due to gravity. He found that while objects of different masses, but of the same material, fell at identical rates, those of different densities did not, and correctly reasoned that this was due to air resistance.

In 1651, Riccioli published his *Almagestum Novum* or New Almagest, an encyclopaedic work of 1,500 pages. This became a standard reference work for many years.

Giovanni Battista Riccioli, as pictured in a 17th century book. (Wikimedia Commons)

In line with the Catholic Church at that time, Riccioli rejected Copernicanism. Rather than the Ptolemaic system, he favoured Tycho Brahe's bizarre "hybrid" model, with the planets orbiting the Sun, while the Sun orbited the Earth! In *Almagestum Novum*, he presented a discussion of opposing world views far more detailed than Galileo's famous *Dialogue*, with 49 arguments in favour of heliocentrism and 77 against. Yet he also cast doubt on several of the latter; it has been suggested that he privately *did* support heliocentrism, but his position in the church prevented him doing so openly.

In one of his arguments for geocentrism, he argued that if the Earth was moving, the stars should show annual parallaxes, but they did not. The reality, of course, is that their parallaxes are far too small to be detected with instruments of that era. In another, he predicted what is now called the Coriolis Effect, whereby the Earth's surface moves eastward with different speeds at different latitudes. He argued that if the Earth rotated, this effect would exist, and would affect the trajectories of cannonballs, but this was not observed. Again, the effect is negligible over the ranges of cannons in his time – but it very much has to be accounted for by modern day artillery!

Riccioli is perhaps best known for introducing the system of lunar nomenclature which is still in use today. He did so in *Almagestum Novum*, using a map of the Moon compiled by his colleague Francesco Grimaldi (1618–1663). Mountain ranges were simply named after those of the Earth, giving us the lunar Alps, Apennines, Caucasus, etc. But the Earth's greatest ranges, the Himalayas and Andes, are missing, as they were not yet known to 17th-century Europeans!

Earlier observers had mistaken the Moon's dark flat plains for water, and named them *maria*, or seas. By 1651, it was well known that they were no such thing, but Riccioli still named them as such. He gave them fanciful Latin names, which are still used as formal names today. Several were named after weather phenomena, due to a mistaken belief that the Moon influenced the Earth's weather. As well as such pleasant-sounding names as the Sea of Tranquillity and Sea of Serenity, he also invented some far less pleasant ones, such as Palus Epidemiarum (the Marsh of Epidemics) and Palus Putredinis (Marsh of Putridity) which are now rarely used, though they still officially exist.

For the craters, of course, he came up with a far better idea. He named them in honour of famous scientists, and the great Greek philosophers, of whom he was an admirer. This tradition has continued to the present day, with many more craters named for later scientists. He greatly admired Copernicus and Kepler, despite the church considering them heretics, so he gave them very prominent craters – but he named the most prominent of all for Tycho Brahe, the last "great astronomer" to hold onto geocentrism.

At first glance, his distribution of crater names appears quite random – but there was in fact some order to it. He divided the Moon into eight 45° sectors or "octants", and populated each one with people of a particular historical period or philosophical tradition.

Riccioli was not exactly impartial; he named a disproportionate number of craters for his fellow Jesuits – though in fairness, they were prominent scientists of their time. Nor did modesty prevent him naming one for himself, and one for his friend Grimaldi. He placed those two names in the same region as Copernicus and Kepler, while the other Jesuits were placed near Tycho – perhaps lending credence to the idea that he and Grimaldi were really "closet Copernicans"!

Riccioli died in Bologna on 25 June 1671.

Sir John Herschel (1792–1871)

This year sees the 150th anniversary of the death of Sir John Herschel, astronomer and scientific polymath.

John Frederick William Herschel was born in Slough on 7 March 1792, the son of the great observer Sir William Herschel. He was educated at Eton College, then St. John's College, Cambridge, where he graduated in 1813 as Senior Wrangler, the highest scoring mathematics graduate.

He shared his father's passion for astronomy, and built an 18-inch reflector in 1816. He was one of the founders of the Royal Astronomical Society in 1820, which elected his father Sir William as

Sir John Herschel at age 75, photographed by Julia Margaret Cameron. (Metropolitan Museum of Art / Wikipedia)

its first President. John himself later served three terms as President.

Herschel proposed the names, which became accepted, of seven of the satellites of Saturn, and of the first four known ones of Uranus – two of the latter having been discovered by his father. Strangely, for those of Uranus, he broke away from the tradition of using names from Roman mythology, and named them after Shakespeare characters. The same convention has been followed for those discovered more recently.

In 1831, he published *A Preliminary Discourse on the Study of Natural Philosophy*, which was an important contribution to the philosophy of science. This described what we now know as the modern scientific method – the iterative process of formulating a theory and testing it by observation and / or experiment. He was knighted in that same year by King William IV.

Between 1833 and 1838, Herschel went to South Africa, and established an observatory near the Cape of Good Hope. There he catalogued the southern sky, to extend his father's previous work to the part of the sky not visible from Britain. He also carried out a variety of other studies, including an extensive study of botany in collaboration with his wife Margaret. On his return, he was made First Baronet of Slough.

In 1864, he published his *General Catalogue of Nebulae and Clusters*, a compilation of both his own work and that of his father.

Herschel was also an early pioneer of photography, and in fact invented the word in 1839. He also discovered the use of sodium thiosulphate, or "hypo", as a fixer.

During his time in South Africa, Herschel was the unwitting subject of the "Great Moon Hoax". In 1835, the *New York Sun* published a series of satirical articles, claiming that Sir John, one of the best known astronomers of the time, using an imaginary "immense telescope", had observed animals on the Moon, including a race of bat-winged humanoids! The intention was probably to ridicule some almost as fanciful claims made by a few eccentric astronomers, such as Franz von Paula Gruithuisen (1774–1852), who claimed to have discovered lunar buildings and cities.

When Herschel himself learned of the hoax on his return to England, he was initially amused, saying that his real discoveries were far less exciting. However, he soon became irritated at having to answer questions from people who actually believed it!

Sir John died at Collingwood, Kent on 11 May 1871. He is buried in Westminster Abbey, alongside such luminaries as Sir Isaac Newton and Charles Darwin.

Sherburne Wesley Burnham (1838–1921)

March sees the centenary of the death of Sherburne Wesley Burnham, an American astronomer best known for his work on double stars.

Burnham was born in Thetford, Vermont on 12 December 1838. He had no advanced education; his knowledge of astronomy was self-taught. In 1858, he became a newspaper reporter, which remained his career for over 40 years. During the Civil War, he was a reporter for the Union Army in New Orleans. It was there that he bought a book which inspired his interest in astronomy.

After the War, he moved to Chicago, where he worked as a court reporter for over 20 years. He became an accomplished amateur astronomer, primarily concerning himself with double stars. Around 1870, he obtained a 6-inch refractor by Alvan Clark, one of the finest American telescope makers of the era.

In the mid-19th century, it was believed that all the binary stars visible to the instruments of the day had been discovered – many of them catalogued by F.

G. W. Struve and his son Otto in Russia. But Burnham, using his 6-inch refractor, discovered 451 new ones between 1872 and 1877. He must have had especially keen eyesight, as he described many doubles as "easy", which other observers had reported as difficult with larger telescopes. In 1874, he published his *General Catalogue of Double Stars*, which remains a standard reference work to this day.

Sherburne Wesley Burnham, by an unknown photographer. (Wikimedia Commons)

The quality of his work led to him being invited to work as a full-time astronomer at Lick Observatory in California from 1888 to 1892, and later at Yerkes Observatory in Wisconsin between 1897 and 1914, where he had access to the world's largest refractor. In 1906, he published the *Burnham Double Star Catalogue*, which listed 13,665 binaries. He is credited with the discovery of 1,340 double stars.

Burnham was awarded the Gold Medal of the Royal Astronomical Society in 1894, and the Lalande Prize of the French Academy of Sciences a decade later. He died at Chicago, Illinois on 11 March 1921. A lunar crater and an asteroid, 834 Burnhamia, are named in his honour.

Henrietta Swan Leavitt (1868–1921)

This year also sees the 100th anniversary of the death, at Cambridge, Massachusetts on 12 December 1921, of **Henrietta Swan Leavitt**, the discoverer of the period-luminosity relationship for Cepheid variable stars. Her life and achievements are commemorated in the article *Henrietta Swan Leavitt and Her Work* by David M. Harland elsewhere in this Yearbook.

Soyuz 11

This year sees the 50th anniversary of a tragic event in the history of spaceflight – the deaths of three cosmonauts in the Soyuz 11 accident.

Soyuz 11 was launched on 6 June 1971. Following a failed attempt by Soyuz 10 a few weeks earlier, its crew were the first to occupy Salyut 1, the world's first space station (see the April Monthly Sky Notes article *Salyut 1: The First Space Station* elsewhere in this Yearbook). The Commander was 43-year-old Lt-Col. Giorgi Timofeyevich Dobrovolsky of the Soviet Air Force; his civilian companions were Vladislav Nikolayevich Volkov, 35, and Viktor Ivanovich Patsayev, who would celebrate

The crew of Soyuz 11 Giorgi Dobrovolsky, Vladislav Volkov and Viktor Patsayev. (USSR Post/Wikimedia Commons)

his 39th birthday aboard Salyut. Volkov was making his second spaceflight, the other two their first. They were in fact the backup crew for the mission; they had been substituted because one of the prime crew had a medical problem. The Commander of the prime crew had been none other than Alexei Leonov, who had made the first spacewalk six years earlier.

The crew set what was then an endurance record of 23 days 18 hours in space – all but about one day aboard Salyut 1. They performed astronomical observations, studied Earth resources and weather, and carried out biological experiments, including growing crops in zero gravity. They were also subjected to extensive biomedical studies, wearing elastic restraining suits for exercise.

But what had been a highly successful mission was to end in tragedy. On 29 June, Soyuz 11 separated from the station to begin its return to Earth. Shortly before re-entry, the crew's radio transmissions ceased. The re-entry and landing procedures were performed automatically – but the recovery team found the crew dead in their seats.

Twelve days later, the cause of their death was announced as "depressurisation" – but the full explanation was not released until three years later. The explosive bolts which separated the re-entry capsule from the orbital module shook open an exhaust valve in the cabin. The outrushing air disoriented the craft; this was corrected by the automatic control thrusters, but confused the crew and delayed their reaction to the open valve. They could probably not have reacted quickly enough anyway, as it took about 45 seconds for their air supply to leak into space.

Dobrovolsky, Volkov and Patsayev were posthumously awarded their country's highest honour, Hero of the Soviet Union. They were given a full state funeral, and their ashes placed in the Kremlin Wall, as was the custom for recipients of that honour. American astronaut Tom Stafford attended on behalf of NASA, and was a pallbearer. Three lunar craters are now named in their honour.

Had the crew been wearing pressure suits, they would have survived. But the Soyuz re-entry capsule was too cramped to accommodate three men in spacesuits; on eight previous Soyuz missions, as well as Voskhod 1 before them, they had flown without them and got away with it. But Dobrovolsky, Volkov and Patsayev paid the ultimate price.

Following the disaster, the Soviet Union flew no more manned spacecraft for two years. When the programme resumed, it was belatedly decided that all future crews would wear spacesuits during launch and re-entry. This meant that the next thirty Soyuz flights had only two-man crews. Finally, beginning in November 1980, the redesigned Soyuz-T was able to carry three men with spacesuits. Soyuz spacecraft are still in service to this day, being used to ferry crews to and from the International Space Station.

But it had taken the needless deaths of three brave men, to convince their political masters that space is no place for cutting corners.

Mission to Mars
Countdown to Building a Brave New World
It All Starts With a Journey

Martin Braddock

Introduction

The insatiable desire to explore is part of the human psyche and drove the American and Russian space programmes of the 1960s and 1970s, resulting in the first successful Moon landing on 20 July 1969. Since then, the eyes of astronomers and the space agencies have been on travelling to Mars with the futuristic vision of establishing and maintaining a human colony on the Red Planet. With the potential to send astronauts to Mars by 2030, this series of articles will address the considerable scientific technical, medical and philosophical challenges of building a Brave New World in this alien environment.

The first article addresses the major medical challenges astronauts will face in their journey to Mars. Even when the Earth and Mars are at their closest proximity of around 55 to 60 million kilometres, it will take approximately nine months to complete the journey. Although this is longer than the six months most astronauts spend on the International Space Station (ISS), to date over 220 individuals have completed their missions onboard the ISS. This provides us with a unique data source enabling us to learn much from astronaut experience of coping with and managing the environment to best protect the physical and mental health of the first generation of humans to establish this new frontier.

In the Beginning

The best time to travel to Mars is during opposition, which is when Mars is closest to Earth and which occurs approximately every two years. During opposition, Mars can be as close as 55 million kilometres from Earth and many space agencies time their missions to coincide with this orbital alignment to send their spacecraft.

The duration of the journey to Mars is between 150 and 300 days, this depending upon the speed of the craft and alignment of the planets. Mariner 4,

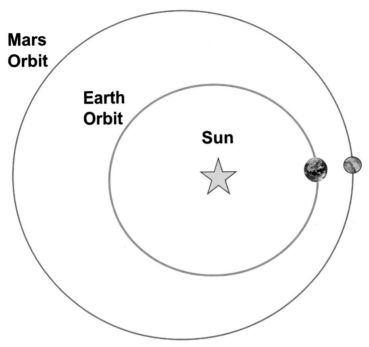

Diagram showing relative positions of Earth and Mars orbits. (Martin Braddock)

launched by the National Aeronautics and Space Administration (NASA) on 28 November 1964, was travelling for 228 days before arriving at the planet on 14 July 1965, whereas Mariner 7, launched on 27 March 1969, reached Mars on 5 August 1969 having required 131 days of space travel. Mariner 9, the first spacecraft to orbit around Mars successfully, launched on 30 May 1971 and arrived at Mars on 13 November 1971 after a voyage lasting 167 days. This journey time of between 150 and 300 days has been maintained for over 50 years of Mars exploration.

Understanding the Environment in Space

Once we have co-ordinated our launch window with a favourable planetary alignment, what might astronauts expect on their typical nine month voyage to Mars? Although space is an alien domain, we have gained great experience for survival and for living and working in an extra-terrestrial environment since Yuri Gagarin became the first human in space in April 1961, completing an Earth

Effect	Current countermeasures	Future countermeasures
Visual disturbances Motion sickness	Vision - no known countermeasure. Pharmacological intervention.	No immediate countermeasure
Sleep quality	Alignment with time exposure to light / dark and adaptation of work schedules	Optimisation and monitoring of cycles, acquisition of data for long (>1 yr) residency in space
Brain function	Adaptation	Study of behaviour for long (>1 yr) residency in space,
Hard and soft tissue loss	Resistance training, nutrition, pharmacological intervention	Optimisation of exercise. Development of tools to predict muscle and bone loss
Healing capacity	Risk-based monitoring to reduce injury, pharmacological intervention	Limited data available on healing rates in humans in space. Translation of animal model to human healing rates in microgravity requires further exploration
Respiratory function	Limit number of EVAs and optimise hyperoxia-inducing regime	Construct 3D human lung models and characterise lung biomarkers under different conditions.
Immune function and infection	Vaccination, screening, pre-mission isolation, diet maintenance	Immunological, microbiome and genomic profiling before mission, monitoring during mission
Cardiovascular function	Resistance training, monitoring and pharmacological intervention	Understanding what level of function is required
Psycho-social function	Crew selection, adequate mental stimulation, contact with Earth, pharmacological intervention	Monitoring complex human behaviour in space missions and other isolation environments. Optimisation of exercise regimes
Gastro-intestinal function	Diet maintenance, pharmacological intervention	Microbiome profiling before mission, monitoring during mission

Table showing current and potential future countermeasures for physical and psychological effects of the environment on astronauts in space. EVA = extra-vehicular activity. (Martin Braddock)

orbit in a flight lasting 108 minutes in the Vostok 1 spacecraft. By 24 September 2019, according to United States Air Force definition, 572 people have been in space with the longest single mission comprising nearly 484 days[1] on board the International Space Station (ISS). During both of these examples and the several hundred other missions undertaken to date, spacecraft ergonomics is constrained by three factors which impact the daily lives of astronauts (Kelly 2017) and present considerable ergonomic challenges for future travel, which include preparing for planetary colonisation.

The first factor is being in an environment of weightlessness or microgravity due to the free fall of space craft in orbit around the Earth or away from the Earth's gravitational attraction in deeper space. The harmful effects of microgravity on human physiology and psychology and potential mitigation measures have been well documented and are shown in the table.

On the ISS all astronauts follow mandatory rigorous exercise regimes, which in part are able to counteract the negative effects of microgravity on, for example, soft and hard tissue wasting (eg Braddock 2017, 2018). Nevertheless, despite

---

1. For further information on who is currently in space and interesting facts on previous astronaut/cosmonaut missions visit **www.worldspaceflight.com/bios/stats.php**

adherence to exercise regimens, current and future availability of medicines (Braddock 2019, Blue 2019a) and compliance with taking pharmaceutical agents known to counteract bone loss on Earth, data from historical ISS missions (Nagaraja & Risin 2013) or ISS and Mir space station missions (Sibonga et al 2015) show that astronauts lost up to 1.6% and 1% of bone and lean muscle mass per month respectively (Nagraja & Risin 2013) or up to 22% of bone mass over the mission (Sibonga et al 2015). After a typical mission of six months duration on the ISS, astronauts are unable to walk unaided on return to Earth. In addition, the environment of microgravity has negative effects on the cardiovascular, central nervous and respiratory systems, weakens the immune response and may affect vision in some astronauts (Braddock 2017). Such is the requirement to find countermeasures for the effects of microgravity, NASA with input from space agencies from Europe and Japan and leading worldwide academics have developed a roadmap for artificial gravity (AG) research (Clement 2017).

The second factor which imposes a more serious challenge to space travel and certainly to Mars colonisation is both astronaut and medicine exposure to radiation (Chancellor et al 2014, Cucinotta 2014, Blue et al 2019b). Unlike on Earth, where the magnetosphere and Van Allen Belts form protective layers from the harmful effects of galactic cosmic radiation (GCR), in space no such protection is available. Astronauts are exposed to GCR which is comprised of protons, helium alphas and heavy ions with very high energies. Current shielding is ineffective as the ions easily penetrate spacecraft and may produce secondary radiation such as X-rays and neutrons. GCR is radiation of high linear energy transfer (LET) which means radiation is deposited in a non-uniform manner along a track. Astronauts also experience solar particle events, which comprise electrons, protons and alpha particles of less energy than GCR, which can be shielded and have low LET being deposited in a uniform manner along the track.

Despite many decades of knowledge of the effects of ionising radiation on living systems on Earth from studies carried out in peacetime and as a consequence of warfare, there is a lack of precision in predicting the health risk to astronauts on both short (< 3 months) and long (> 3 months) term missions, and recent studies have attempted to address the limitations (Chancellor et al 2018, Blue et al 2019b). It has been recommended that astronauts are limited to career lifetime radiation exposures that would induce no more than a 3% risk of exposure-induced death and this translates into a radiation dose

Mars Explorers Wanted! Do you want to help build a new life and society on Mars? Do you have a sense of adventure and a positive attitude? If so, apply to join us! (NASA/KSC)

dependent upon astronaut sex and age. Although estimates vary, for a Mars mission of 975 days which includes 536 days in transit, astronauts would receive a radiation dose of approximately 2 Sievert (Sv) (Sion 2011). This compares with background radiation exposure on Earth of between 1 and 2 milliSv per year and a recommended 2.5 and 1.75 Sv lifetime exposures for a 35-year-old male and a 35-year-old female astronaut respectively. It is clear therefore that the current advisory limit for astronaut exposure to radiation will be challenged and may limit mission duration and would very likely be exceeded on longer missions[2] and in the establishment of a colony on Mars unless new mitigation measures are developed.

2. NASA Likely to Break Radiation Rules to Go to Mars **www.pbs.org/wgbh/nova/article/nasa-mars-radiation-rule**

A multi-disciplinary team led by the NASA Ames Research Centre has published a roadmap for an integrated approach to understanding how radiation resistance could be conferred upon astronauts (Cortese et al 2018), one property which falls under the category of human enhancement. The team outlines future research directions which includes upregulation of endogenous DNA repair and radioprotection mechanisms and isotopic substitution of organic molecules. In parallel, advances in drug discovery have isolated molecular targets which may confer drug resistance, which is may be amenable to both astronaut radioprotection and for the successful subversion of radio protective mechanisms for the treatment of patients with cancer.

The third factor which provides both a physical but especially a psychological challenge for astronauts is to live and work for long periods of time in a confined environment (Kanas 2010, 2011). In space, there is no 'going outside'

A glimpse of our new home – different and beautiful! (NASA)

for a break and although the ISS provides a spacious habitat for the crew,[3] there is the inevitable monotony of being in the same location for many months and with the same people. This may lead to mood changes and periods of anxiety or depression which could affect personal performance and that of the crew in executing the mission. The effects of working in a confined environment, in some cases much smaller than the ISS as in the Apollo missions of the 1960s and 1970s, have led to the development of numerous terrestrial-based analogues (eg Kanas et al 2011, Pagel & Chouker 2016), countermeasures (Sandal 2001) and mitigation strategies which include pharmaceutical intervention with anti-depressants (Stingl 2015), securing routes to enable expression of state of mind[4] and adequate rest.[5]

In summary, we believe we have the skills and ability to get a spacecraft to Mars carrying a human cargo. The next article in this series will address what protection could be provided to astronauts to safeguard their physical and mental well-being. The roles of emerging scientific innovation in extending human longevity, including the potential of human enhancements, are exciting developments which are on our event horizon!

References

Blue, R. S., Bayuse, T. M., Daniels, V. R. et al (2019a). 'Supplying a pharmacy for NASA exploration spaceflight: challenges and current understanding', *npj Microgravity* 5: 14.

Blue, R. S., Chancellor, J. C., Antonsen, E.L. et al (2019b). 'Limitations in predicting radiation-induced pharmaceutical instability during long-duration spaceflight', *npj Microgravity* 5: 15.

Braddock, M (2017). 'Ergonomic challenges for astronauts during space travel and the need for space medicine', *Journal of Ergonomics* 7: 1–10.

Braddock, M (2018). 'Exercise and ergonomics on the International Space Station and Orion spacecraft', *Journal of Ergonomics Research* 1:2 doi: 10.4172/JEOR.1000104.

---

3. International Space Station Facts and Figures **nasa.gov/feature/facts-and-figures**

4. Current Psychological Support for US Astronauts on the International Space Station **ntrs.nasa.gov/archive/nasa/casi.ntrs.nasa.gov/20070035862.pdf**

5. Behavioural Issues Associated with Isolation and Confinement: Review and Analysis of Astronaut Journals **www.nasa.gov/mission_pages/station/research/experiments/explorer/Investigation.html?#id=964**

Braddock, M (2019). 'From target identification to drug development in space: using the microgravity assist', *Current Drug Discovery Technologies* 16: 1–11.

Chancellor, J.C., Scott, G.B.J., & Sutton, J.P. (2014). 'Space radiation: the number one risk to astronaut health beyond low Earth orbit', *Life* 4: 491–510.

Chancellor, J.C., Blue, R.S., Cangel, K.A., Aunon-Chancellor, S.M., Rubins, K.H., Katzgraber, H.G., & Kennedy, A.R. (2018). 'Limitation in predicting the space radiation health risk for exploration astronauts', *npj Microgravity* 4:8 doi:10.1038/s41526-018-0043-2.

Clement, G. (2017). 'International roadmap for artificial gravity research', *npj Microgravity* 3:29 doi:10.1038/s41526-017-0034-8.

Cortese, F., Klokov, D., Osipov, A. et al (2018). 'Vive la radioresistance!: converging research in radiobiology and bio gerontology to enhance human radio resistance for deep space exploration and colonisation', *Oncotarget* 9: doi: 10.18632/oncotarget 24461.

Cucinotta, F.A. (2014). 'Space radiation risks for astronauts on multiple international space station missions', *PLOS ONE* 9:396099.

Kanas, N. (2010). 'Expedition to Mars: psychological, interpersonal and psychiatric issues', *Journal of Cosmology* 12: 3741–3747.

Kanas, N. (2011). 'From Earth's orbit to the outer planets and beyond. Psychological issues in space', *Acta Astronautica* 68: 576–581.

Kelly, S. (2017). Endurance, Knopf Publishing Group.

Nagaraja, M.P., & Risin D. (2013). 'The current state of bone loss research: Data from spaceflight and microgravity simulators'. *Journal of Cellular Biochemistry* 114: 1001–1008.

Pagel, J.I., & Chouker A. (2016). 'Effects of isolation and confinement on humans implications for manned space explorations', *Journal of Applied Physiology* 120: 1449–1457.

Sandal, G.M. (2001). 'Psycho-social issues in space: future challenges', *Gravitational and Space Research* 14: 47–54.

Sibonga, J.D., Spector, E.R., Johnston, S.L., & Tarver, W.J. (2015). 'Evaluating bone loss on ISS astronauts', *Aerospace Medicine & Human Performance* 86: A38–A44.

Sion, N. (2011). 'Can astronauts survive radiation on prolonged space missions?', *Bulletin of the Canadian Radiation Protection Association* 31: 20–26.

Stingl, J.C., Welker, S., Hartmann, G., Damann, V., & Gerzer, R. (2015). 'Where failure is not an option – personalised medicine in astronauts', *PLOS ONE* 10:e0140764.

Male Family Mentors for Women in Astronomy En'hedu'anna to Eimmart

Mary McIntyre

Much has been written over the years about trailblazing women in astronomy in the eighteenth and nineteenth centuries. Whilst their impact was huge, they were not the first; women in astronomy can be traced back centuries before. Women astronomers sometimes had the support of a male partner or family member. This is part one of a three-part article looking at some of these partnerships and, although not an exhaustive list, it features some very inspirational female astronomers …

Throughout history there has been a huge disparity between the number of men and women working in astronomy. There was a general belief that women were not mentally capable of being astronomers. The eighteenth century writer Jean-Jacques Rousseau commented: "Women do not possess sufficient precision and attention to succeed at the exact sciences." They were often forbidden from using libraries, and they were not permitted to attend university lectures until the 1800s, because male classicists feared women would be corrupted by the "shocking stories and filthy words" from classical authors.

However, despite a lack of formal education, many women were participating in science and astronomy. In Germany between 1650 and 1710 around 14% of astronomical scientists were women. Later, when women did attend lectures, they were unable to graduate. Physics and astronomy were taught in Cheltenham Ladies' College by the 1860s, but it was not until colleges such as Bedford College in London and St. Hilda's in Oxford were founded that women could graduate. Even then, to get a degree, a lady needed to be clever, determined, and have a wealthy, understanding family who could accept she may want a life beyond domesticity. In addition to the struggle getting a good education, women were not admitted as full Fellows of the Royal Astronomical Society until 1916. Finally if a woman had persevered and got herself a good education and a job working in astronomy, she often had to give up work if she got married.

Founded in 1849 by English social reformer and philanthropist Elizabeth Jesser Reid, Bedford College for Women, London was one of the first colleges in the United Kingdom to provide higher education for women, and to allow women to graduate. (Wikimedia Commons)

So what did women do to further their astronomy education with all of these obstacles in their way? In many cases, they had a mentor in the form of a male family member, and were often taught privately at home by their father, grandfather, brother, uncle, or husband. Women would assist these men, but often worked hard on their own observations. There are many cases throughout history where women made an astronomical discovery, but the male mentor was given credit for it. Sometimes this was malicious, other times it was because women were simply not permitted to report a discovery. This situation was still happening in recent years (Jocelyn Bell Burnell discovered pulsars in 1967, although it was her supervisor who got the credit). This was not unique to astronomy; it was repeated many times across the sciences.

Women who practiced astronomy can be traced back millennia. Around 2,500 BC, female Sumerian scribes were responsible for teaching. One female poet from this period is En'hedu'anna (c. 2,300 BC). She was given the role of chief astronomer-priestess by her father, Sargon of Akkad. She was a learned, talented and powerful lady, managing the great temple complex in her city.

The lady credited as the first female astronomer in the western world is Aglaonice, also known as Aganice of Thessaly. Championed by her father, she worked on Saros cycles and accurately predicted lunar eclipses. She also ran a school for female astronomers, which the male scientific community referred to as the "witches of Thessaly". They were active from the first to third centuries BC.

Hypatia of Alexandria, born c. 350–370 AD, was a philosopher, mathematician and astronomer from Egypt. She was the daughter of a mathematician and taught

Portrait of Hypatia of Alexandria drawn in 1908 by the French-born American portrait painter Jules Maurice Gaspard (1862–1919) and featured in *Little Journeys to the Homes of the Great* by American writer Elbert Hubbard (Wikimedia Commons/Jules Maurice Gaspard)

This illumination from the *Liber Scivias* shows Hildegard of Bingen receiving a vision and dictating to her scribe and secretary. Hildegard was famous for having visions, which she recorded and made into three volumes of visionary theology. (Wikimedia Commons/Scivias)

A portrait of Sophia Brahe from 1602. Despite the large age gap, she was passionate about astronomy and was helping Tycho with his observations by the time she was ten years old. (Wikimedia Commons)

philosophy and astronomy, and is known to have built astrolabes. Hypatia is probably the first female mathematician whose life was well recorded. She was held in high regard by everybody and the empire was shocked when she was murdered in 415.

The period that followed was a dark time for science in general, but there was also a cultural shift which meant women were excluded from social and cultural life. Despite this, there were still women who were involved in science during the Dark Ages, one exceptionally talented lady being Hildegard of Bingen (1098–1179). Another worthy of note is the Islamic astronomer Fátima of Madrid (tenth or eleventh century). Fátima was the daughter of an astronomer and she worked with him on astronomical and mathematical investigations.

These included the correction of astronomical tables, preparing calendars and calculating the positions of the Sun, Moon and planets. In addition, Fátima wrote several works known as *Corrections of Fátima*. We know that women were still active within the science community during the Dark Ages because in the translation of Euclid's Elements c. 1310, there is an illustration in which women are depicted teaching geometry.

When we hear about brother and sister astronomy partnerships we immediately think of the Herschels, but there was another famous astronomer pre-dating them who had assistance from his sister. This was the Danish nobleman Tycho Brahe whose youngest sibling Sophia (1556–1643) had developed a love of the stars during early childhood. Sophia was over a decade younger than Tycho, yet by the age of ten she was helping him with his observations. They both had a passion for science; one that their family disapproved of, feeling it was an inappropriate pastime for people of nobility. Although Sophia wanted to study at university, her sex prevented it. However, she and Tycho were united by their love of astronomy and he helped to educate her in many different subjects. Sophia worked with Tycho at his observatory, calculating eclipses and cometary paths, and she was the first person to accurately measure the position of the planets. Sophia was forced to marry by her parents so she had to give up work. Following the death of her husband, she continued to help Tycho in his observatory as well as furthering her education in other subjects.

Another remarkable female astronomer was Maria Cunitz, born sometime after 1604 but before 1610 in Silesia (now Poland). Her father was a medical doctor and he educated her in many languages, art, music and poetry; all subjects considered suitable for a lady. In addition she was taught subjects not considered suitable, including mathematics, astronomy, medicine and history, by a doctor and amateur astronomer called Elias von Löwen. There are records of planetary observations made by the two of them in 1627 and 1628. Maria and Elias married in 1630 and he actively encouraged her enthusiastic interest in astronomy. In 1727, Johann Kaspar Elberti wrote that: "[Cunitz] was so deeply engaged in astronomical speculation that she neglected her household. The daylight hours she spent, for the most part, in bed [...] because she had tired herself from watching the stars at night." Her observational skills were hindered by her lack of access to good quality instruments so instead she applied her mathematical skills to astronomy. She published her mathematical work *Urania*

A memorial statue of Maria Cunitz in Swidnica, Poland, photographed in June 2014. Maria was taught astronomy and mathematics by German mathematician Elias von Löwen, to whom she was married in 1630. (Wikimedia Commons/Piotrus)

Propitia in 1650, in which she corrected some errors in Kepler's tables. Maria actively corresponded with other leading astronomers and mathematicians, although because this went against convention, most of the correspondence was addressed to and from her husband. All of their astronomy observations and papers were destroyed by fire in 1656.

It was in 1639 that Johannes Hevelius of Gdansk, Poland, began to devote himself to astronomy full time, using his family wealth to build a private observatory. Hevelius was both well respected and well connected, and in 1664 was made a Fellow of the Royal Society in London. On 17 January 1647, a girl called Elisabetha Koopman was also born in Gdansk. Elisabetha was fascinated by the night sky and when she was still a young child, she approached Hevelius, asking him to teach her. He promised that he would show her the "wonders of the heavens" when she was older. She approached him again when she was 15 years old and he took her under his wing. There is a beautiful quote written by E. F. MacPike describing Elisabetha observing the Moon: "And when in the star-lit night she followed with enraptured gaze and beating heart, through his giant telescope, the shinning full moon on her silent path, she exclaimed with enthusiasm: 'To remain and gaze here always, to be allowed to explore and proclaim with you the wonder of the heavens, that would make me perfectly happy!' And the worthy man felt that it might make him happy too." In 1663, 52-year old Johannes took 16-year old Elisabetha as his second wife and she became his astronomy assistant. Elisabetha performed her own observations

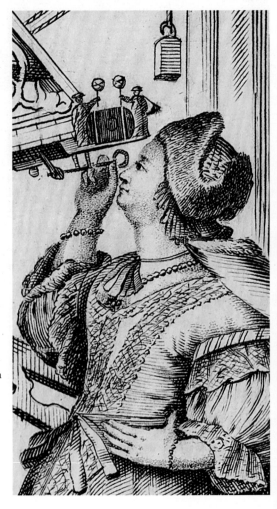

This engraving by Isaak Saal (after a drawing by Baroque artist Andreas Stech and taken from the first part of *Machina Coelestis*, written by Johannes Hevelius and published in 1673) depicts Elisabetha Hevelius observing the sky with a brass octant. (Wikimedia Commons/ Isaak Saal/Andreas Stech)

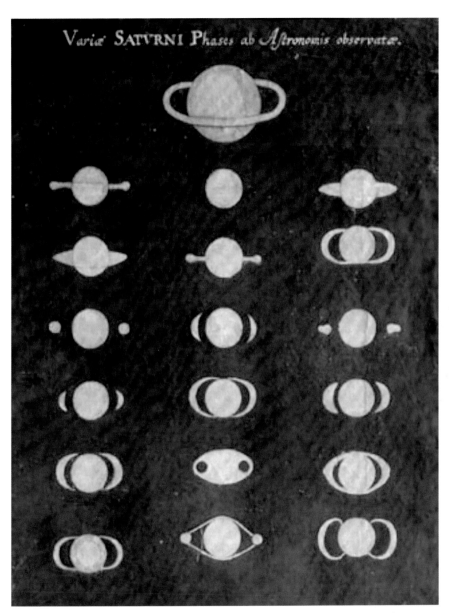

Aspect of Saturn sketch by Maria Clara Eimmart. Maria's father was a painter and engraver and he taught her how to draw and engrave. She was a prolific sketcher and is best known for her exact astronomical illustrations created with pale pastels on dark blue card. (Wikimedia Commons/Maria Clara Eimmart)

and undertook important mathematical calculations. Following the death of her husband, she prepared three of his books for publication. Much of the content of those books is said to be due to Elisabetha's work. Most of her correspondence and records were lost in a fire that destroyed their home and observatory.

Maria Margarethe Winklemann was born in Germany in 1670. Her father was an advocate of education for women and he taught her from an early age. When her father died, her uncle took over the role of teacher. Like many before her, Maria showed interest in astronomy from a very young age and she became an apprentice of the self-taught astronomer Christopher Arnold. It was through Arnold that she met one of Germany's best known astronomers, Gottfried Kirch. They married in 1692 and she became his assistant. Together they produced calendars and almanacs, predicted eclipses, and the positions of the Sun, Moon and planets. In 1702 Maria became the first woman to discover a new comet (although it was credited to Gottfried). Her skill and accomplishments were widely recognised and she was held in high esteem, so it is absolutely shocking that when her husband died in 1710, the academy cut her loose. She found work in some other private observatories, but conditions meant she was forced to abandon astronomy altogether.

Our third "Maria" is Maria Clara Eimmart. Born in 1676 in Nuremberg, she was the daughter of amateur astronomer Georg Christoph Eimmart, who was a painter and engraver. This was a lucrative profession and Georg was able to invest in astronomy instruments and fund the building of a private observatory. Maria trained as his apprentice, and in addition to teaching her how to draw and engrave, he also educated her in French, Latin, mathematics and astronomy. She became well known for her engravings depicting lunar phases. Maria married Johann Heinrich Muller in 1706. Johann was a former student of her father, a physics teacher and director of their observatory. Maria's passion for astronomy inspired him to become a well-respected amateur astronomer himself. Maria assisted him with observations and she became a prolific sketcher at the eyepiece. Her telescopic illustrations depicted the ever changing lunar landscape, planets, comets, eclipses and atmospheric optical effects, and her pastel sketches on dark blue paper are truly gorgeous. Maria died in Nuremberg in 1707, just a year after her marriage.

Perhaps the best known brother-sister astronomy partnership was that of William and Caroline Herschel. We will come to them in part two of this article in the forthcoming *Yearbook of Astronomy 2023*.

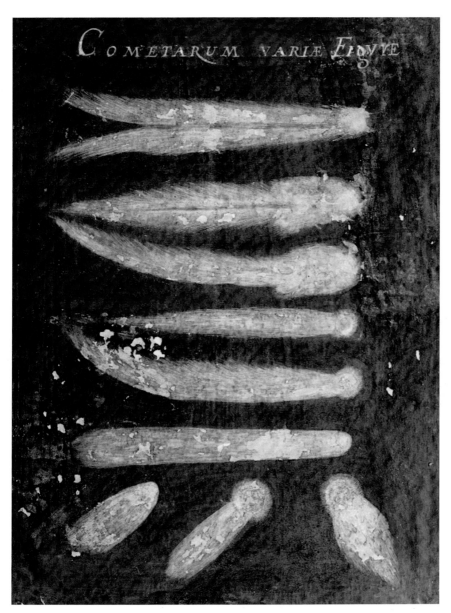

Two other examples of Maria's work are shown here. Her sketches were exquisitely detailed and this is evident in the one depicting nine comets. Between 1693 and 1698 she created over 350 drawings of the phases of the Moon. These were so accurate that they were used to create a new lunar map. (Wikimedia Commons/Maria Clara Eimmart)

Henrietta Swan Leavitt and Her Work

David M. Harland

This article marks the centenary of the death of Henrietta Swan Leavitt, whose discovery of the period-luminosity relationship for a given type of variable star later revealed the vast size of the universe.

* * *

Henrietta Swan Leavitt was born in Lancaster, Massachusetts, on 4 July 1868, inheriting her forenames from her mother, Henrietta Swan Kendrick. She was the oldest of seven children, two of whom died very young. Her father, George Roswell Leavitt, was a graduate of Williams College, had a doctorate in divinity from Andover Theological Seminary, and was a Congregationalist minister. The family traced its ancestry back four centuries to Yorkshire in England, and was regarded as what was then known as "good Puritan stock".

When Leavitt was a teenager the family moved to Cleveland, Ohio, where her father became pastor of the Plymouth Congregational Church.

Her parents believed fervently in education, so when Leavitt reached 17 she was enrolled at the Oberlin Conservatory of Music for two years' of undergraduate studies. In addition to the liberal arts, she excelled in mathematics.

In 1888, now aged 20, Leavitt returned to Massachusetts and entered the Society for Collegiate Instruction of Women in Cambridge, created in 1879 as the Harvard Annex to enable women to

Henrietta Swan Leavitt. (Harvard College Observatory)

receive tuition from the teaching staff at Harvard, which at that time was all male. (It would become Radcliffe College in 1894.)

In her fourth year, Leavitt expanded her curriculum in the arts and humanities to include a course on astronomy that was taught by staff at the university's observatory.

In fact, the Harvard College Observatory (HCO), was the oldest such facility in the United States. When Harvard began to consider opening an observatory, William Cranch Bond, a prominent Boston clockmaker, was dispatched to Europe to inspect similar institutions there. On his return, Bond submitted his favourable report and when the observatory was established in 1844 he was appointed as director. After installing a refractor with a lens diameter of 15 inches, the largest such instrument in the country and therefore known as the 'Great Refractor', Bond set about applying photography to astronomy because it enabled the cataloguing of celestial objects on a scale far surpassing what could be achieved by visual observations.

After a succession of directors, in 1877 the HCO passed to Harvard alumnus Edward Charles Pickering who was then teaching physics at the Massachusetts Institute of Technology in Cambridge. Dissatisfied with the performance of male assistants in analysing photographic plates, Pickering gave the task to his Scottish housemaid, Williamina Fleming, because he thought a woman would be more fastidious. When this was established, he set about hiring female assistants called 'computers' to assist with the growing task of classifying stellar spectra provided by a telescope fitted with an objective prism.

When Leavitt graduated in 1892, just short of her 24th birthday, the fact that Harvard was not able to award degrees to women meant she received a certificate that was the equivalent of a BA degree. She decided to stay in Cambridge to earn graduate credits while helping out as a volunteer at the observatory.

Edward Charles Pickering.
(Harvard College Observatory)

In 1895 Pickering delightedly accepted Leavitt as an unpaid assistant at the observatory because she had scored very highly in the astronomy courses. He assigned her to stellar photometry, gauging the brightness of stars by the size of the dark spots on 'negative' photographic plates, with brighter stars producing larger spots. The initial study focused on the northernmost stars, with the aim of defining a sequence of stars at a range of magnitudes to provide references for comparison with stars in other parts of the sky. The aim was to test a correlation between the visual and photographic measurements, in the hope that in future it would be possible to use the photographic technique instead of having astronomers consume a lot of valuable telescope time by visually estimating the brightness of the stars individually.

As part of this photometric work, Leavitt paid particular attention to stars whose brightness changed over time. Finding 'variables' involved comparing plates of the same field taken at different times.

In 1896 Leavitt wrote up her north-polar photometry (including the variable stars in that zone) and then sailed for Europe, where she spent two years. Where she went and what she did is not recorded, because she left no diaries and no boxes of letters.

Back in the USA, Leavitt joined her family in Wisconsin, where her father had a new ministry, and she took a position as an art assistant at Beloit College.

On finding herself eager to resume astronomical work, in May 1902 Leavitt wrote to Pickering seeking his advice. He wrote back promptly, offering her a full-time position. In recognition of her excellent work as a volunteer he offered her a salary of 30 cents per hour rather than the standard 25 cents. She arrived in August and initially worked part time until becoming a full time member of staff the following year. Her title was 'reader' because Pickering regarded the photographic plates of the sky as the equivalent of books in a library, and it was her job to analyse them.

Her first target was the Orion Nebula (M42), for which the HCO had a decade's worth of plates taken by a variety of telescopes. To estimate the brightness of stars on a negative plate, she used a rectangular glass palette that was about 1 inch by 3 inches, framed in metal and mounted on a long, thin handle. It resembled a fly-swatter, but being "too small to do a fly much damage" she called it her "fly-spanker". By measuring the brightness of all the stars in the target area on a succession of plates she identified and monitored variables

The outstation at Arequipa in the Peruvian Andes, circa 1900, with El Misti in the background. (Harvard College Observatory)

associated with the nebula. But measuring the magnitudes of large numbers of stars in a given field on a series of plates to find those that were varying was a time-consuming process, so Pickering provided her with a 'positive' plate as a reference. By superimposing a negative over the positive, the dark and bright spots of stars that were not changing cancelled out and those that were varying literally shone. The number of discoveries rapidly increased.

In 1891 Pickering had established an outstation at Arequipa, high in the Peruvian Andes, to extend the HCO photographic coverage into the southern sky. Two years later, Solon Irving Bailey was assigned to this pioneering work.

Solon Irving Bailey. (National Academy of Sciences)

On being appointed director of the Cape Observatory in 1834, John Herschel had erected his father's favourite 20-foot reflector and made a comprehensive survey of the southern sky. Four years later, back in London, he began to process his accumulated observations, and in 1847 issued a catalogue which increased to almost 100 the number of dense spherical agglomerations of stars called 'globular clusters'. Surprisingly most of them were located in the constellation of Sagittarius.

Bailey developed a keen interest in globulars driven, in part, by the fact that they were rich in variables that cycled with periods ranging from about 0.2 to 1.0 days with a characteristic light curve that spanned about half a magnitude. These were so distinctive that he coined the name 'cluster variables'.[1] Significantly, because all of the stars in any given globular were about the same distance from us, the fact that the variables were of similar brightness meant that they must be about the same absolute magnitude. Short exposures were required to achieve sufficient time-resolution to trace the light curves of such rapid cycles. To assist with this investigation, Pickering requested that multiple exposures be taken on a single plate, progressively displacing the image in order to make the stars form lines of dots and thus simplify the measurement of the magnitude variations in relative terms.

Bailey was also devoting attention to two patches of nebulosity that resembled the Milky Way but were isolated from it. These had been named in honour of Ferdinand Magellan, who noted their existence in his ship's log when he ventured into the far southern ocean in 1591.

Whereas a gaseous nebula such as M42 contained only a smattering of stars, the Magellanic Clouds were densely populated. In early 1904, Pickering assigned Leavitt the task of analysing the plates of the clouds taken by the 24-inch astrographic refractor at Arequipa, known as the Bruce Telescope because its construction had been funded by the wealthy New Yorker, Miss Catherine Bruce.

1. These variables proved to be similar to the RR Lyraes that are found in the Milky Way system. In fact, the prototype was discovered by Williamina Fleming on plates at the HCO and reported by Pickering in 1900 as being "indistinguishable from cluster-type variables".

Leavitt began by superimposing a fine grid over a plate in order to measure the positions of all the stars in the outer regions of each cloud.[2] The plates were 14 × 17 inches, and the 1 cm spacing of her transparent overlay equated to 10 minutes of arc. Once the celestial coordinates of the grid itself had been precisely determined, she was able to measure the positions of the stars. Then she drew up a photometric sequence of stars of constant brightness in order to be able to measure the variations of others.

The Bruce Telescope built in 1893 by Alvan Clark & Sons was an astrographic refractor with a 24-inch lens. In 1896 it was installed at Arequipa. It is seen here in 1904. (Harvard College Observatory)

Leavitt was soon finding surprisingly large numbers of variables in both clouds, and they were reported as a series of *Harvard College Observatory Circulars*. In a personal letter to Pickering in March 1905 Charles Augustus Young at Princeton University praised her as "a variable-star fiend. One cannot keep up with the roll of new discoveries."

This article will focus on the Small Magellanic Cloud (SMC), because this study led to her most significant astrophysical discovery.

Many of the variables in the SMC became brighter than the fifteenth magnitude at maximum but very few exceeded the thirteenth. Even with the Bruce Telescope, exposures of 2 to 4 hours were necessary. Determining the light curves of variables having rapid periods required exposing as many plates as possible on a given night.

After several years of detailed study, in 1908 Leavitt issued a 24-page report in the *Annals of Harvard College Observatory* in which she gave the coordinates of

2. The Magellanic Clouds are so densely populated that on the Arequipa plates Leavitt was able to resolve individual stars only in their outer regions.

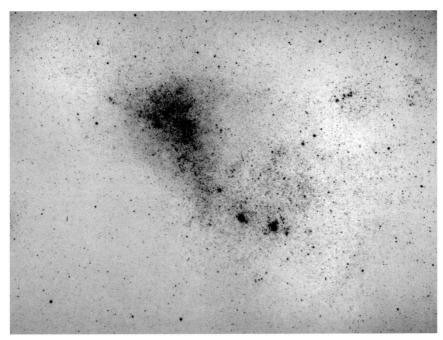

A negative plate of the Small Magellanic Cloud viewed from Arequipa. (Harvard College Observatory)

variables and their maximum/minimum magnitudes.[3] Of those for which she had satisfactory light curves, she drew particular attention to 16 with periods ranging from 1.25 to 127 days, saying:

"The majority of the light curves have a striking resemblance, in form, to those of the cluster variables. As a rule, they are faint during the greater part of the time, the maxima being very brief, while the increase of light usually does not occupy more than from one-sixth to one-tenth of the entire period. It is worthy of notice that the brighter variables have the longer periods. It is also noticeable that those having the longer periods appear to be as regular in their variations as those which pass through their changes in a day or two."

---

3. This paper is '1777 Variables in the Magellanic Clouds', *Annals of Harvard College Observatory*, vol. 60, #4, 1908 (month not specified). It included 15 pages of tables and two plates (one each for the Large and Small Magellanic Clouds).

PERIODS OF VARIABLES IN THE SMALL MAGELLANIC CLOUD.

Harvard No.	Max.	Min.	Range.	Epoch.	Period.	Min. to Max.	Average Dev.	Earliest Observation.	No. Periods.	No. Plates.
					d.	d.				
818	13.6	14.7	1.1	4.0	10.336	1.7	.12	1890	566	44
821	11.2	12.1	0.9	97.	127.	49.	.06	1890	45	89
823	12.2	14.1	1.9	2.9	31.94	3.	.13	1890	184	56
824	11.4	12.8	1.4	4.	65.8	7.	.12	1889	94	83
827	13.4	14.3	0.9	11.6	13.47	6.	.11	1890	448	60
842	14.6	16.1	1.5	2.61	4.2897	0.6	.06	1896	843	26
1374	13.9	15.2	1.3	6.0	8.397	2.	.10	1893	574	42
1400	14.1	14.8	0.7	4.0	6.650	1.	.11	1893	724	42
1425	14.3	15.3	1.0	2.8	4.547	0.8	.09	1893	1042	33
1436	14.8	16.4	1.6	0.02	1.6637	0.3	.10	1893	2859	22
1446	14.8	16.4	1.6	1.38	1.7620	0.3	.09	1896	2052	21
1505	14.8	16.1	1.3	0.02	1.25336	0.2	.10	1896	2335	25
1506	15.1	16.3	1.2	1.08	1.87502	0.3	.09	1896	1560	23
1646	14.4	15.4	1.0	4.30	5.311	0.7	.06	1896	681	24
1649	14.3	15.2	0.9	5.05	5.323	0.7	.10	1893	894	32
1742	14.3	15.5	1.2	0.95	4.9866	0.7	.07	1893	954	28

The 16 variables in the SMC reported by Leavitt in 1908. (Harvard College Observatory)

To reinforce the sheer amount of data underlying this generalisation, she noted that for the variable whose period was 127 days she had 89 observations that spanned 45 maxima, with an average deviation from the light curve of only six hundredths of a magnitude.

In addition to progressive loss of hearing, Leavitt suffered frequent ailments. Toward the end of 1908 she was sufficiently ill to be admitted to a hospital in Boston. She returned to Wisconsin at the start of the next year in order to convalesce with her family. She had hoped to resume work in September, but her illness persisted.

Arequipa continued to supply plates of the SMC during her absence, so when she finally resumed work in May 1910 she checked up on known variables and set about discovering new ones. There was another pause for six months when she returned to Wisconsin after the death of her father in March 1911.

Having found another nine variables in the SMC with the distinctive light curve, giving her a total of 25, she plotted the periods against their limiting magnitudes using a linear scale and obtained a pair of smooth curves. On

PERIODS OF VARIABLE STARS IN THE SMALL MAGELLANIC CLOUD.

H.	Max.	Min.	Epoch.	Period.	Res. $M.$	Res. $m.$	H.	Max.	Min.	Epoch.	Period.	Res. $M.$	Res. $m.$
			d.	d.						d.	d.		
1505	14.8	16.1	0.02	1.25336	−0.6	−0.5	1400	14.1	14.8	4.0	6.650	+0.2	−0.3
1436	14.8	16.4	0.02	1.6637	−0.3	+0.1	*1355*	14.0	14.8	4.8	7.483	+0.2	−0.2
1446	14.8	16.4	1.38	1.7620	−0.3	+0.1	1374	13.9	15.2	6.0	8.397	+0.2	−0.3
1506	15.1	16.3	1.08	1.87502	+0.1	+0.1	818	13.6	14.7	4.0	10.336	0.0	0.0
1413	14.7	15.6	0.35	2.17352	−0.2	−0.5	*1610*	13.4	14.6	11.0	11.645	0.0	0.0
1460	14.4	15.7	0.00	2.913	−0.3	−0.1	*1365*	13.8	14.8	9.6	12.417	+0.4	+0.2
1422	14.7	15.9	0.6	3.501	+0.2	+0.2	*1351*	13.4	14.4	4.0	13.08	+0.1	−0.1
842	14.6	16.1	2.61	4.2897	+0.3	+0.6	827	13.4	14.3	11.6	13.47	+0.1	−0.2
1425	14.3	15.3	2.8	4.547	0.0	−0.1	*822*	13.0	14.6	13.0	16.75	−0.1	+0.3
1742	14.3	15.5	0.95	4.9866	+0.1	+0.2	823	12.2	14.1	2.9	31.94	−0.3	+0.4
1646	14.4	15.4	4.30	5.311	+0.3	+0.1	824	11.4	12.8	4.	65.8	−0.4	−0.2
1649	14.3	15.2	5.05	5.323	+0.2	−0.1	821	11.2	12.1	97.	127.0	−0.1	−0.4
1492	13.8	14.8	0.6	6.2926	−0.2	−0.4							

The 25 variables in the SMC reported by Leavitt in 1912. The entries in italic are the 9 stars added since her first paper. (Harvard College Observatory)

replotting the data using a logarithmic scale she got straight lines.[4] The slope indicated that the logarithm of the period increased by about 0.48 for an increase of one magnitude in brightness. The average difference between the maximum and minimum brightness for bright and faint variables alike was about 1.2 magnitudes.

In March 1912, this period-luminosity relationship was published in a 3-page paper that carried an introduction by Pickering and bore his name.[5]

If the SMC was a distinct system of stars and very far away, she observed: "Since the variables are probably at nearly the same distance from Earth, their periods are apparently associated with their actual emission of light, as determined by their mass, density, and surface brightness."

A study of the physical characteristics of stars required spectroscopy but, as Leavitt lamented: "The faintness of the variables in the Magellanic Clouds seems to preclude the study of their spectra, with our present facilities." Nevertheless,

---

4. There was no plot in the 1908 paper, just a statement that the longer the period, the brighter the apparent magnitude of the star.

5. 'Periods of 25 Variable Stars in the Small Magellanic Cloud', *Harvard College Observatory Circulars*, #173, 1912 (March).

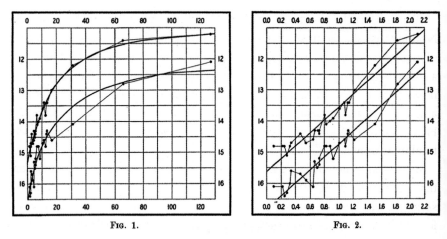

Fig. 1.   Fig. 2.

The period-luminosity relationship discovered by Leavitt, with a logarithmic brightness scale on the right. (Harvard College Observatory)

she pointed out that there were a number of stars with similar light curves elsewhere in the sky, and because these Cepheids were brighter (presumably because they were closer) she recommended a spectroscopic study of them.

Wrapping up her 1912 paper, Leavitt expressed the hope that "the parallaxes of some variables of this [Cepheid] type may be measured" to determine their distances and thus their true luminosities.[6] She was aware that if the relationship could be calibrated then such stars would serve as "standard candles" (her term). Thereafter, the distance to such a star would be able to be determined simply by measuring the period of its light curve. But she was not allowed to follow up the implications of her discovery because Pickering insisted that she push on with the search for variables in the other zones of the sky.

Pickering's attitude reflected his view of the HCO's mission; namely to take and analyse plates, then compile catalogues of data to facilitate research by *other* institutions. Of course, another factor may have been that, despite his appreciation of the skills of his female assistants, they were computers of data rather than astronomers. If Leavitt had been a male postgraduate student, she could easily have been awarded a doctorate for her work on the Magellanic Clouds.

---

6. Note that in her papers Leavitt never actually referred to her variables as being of the Cepheid type.

Ill health struck Leavitt again early in 1913 requiring surgery on her stomach. Nothing that she did in subsequent years was as momentous as the discovery of the period-luminosity relationship. In 1918 she was finally diagnosed with a malignant stomach cancer, and after a long and painful decline she died in hospital on 12 December 1921, aged 53. She was buried in the family plot on a shallow hill at Cambridge Cemetery. During her career as a computer at the HCO, she had discovered some 2,400 variables, which was about half of the total then known.

In 1924 the renowned Swedish mathematician Gösta Mittag-Leffler, unaware that Leavitt was dead, wrote to her expressing his admiration of her work and saying he was "seriously inclined to nominate [her] for the Nobel Prize in Physics for 1926". Unfortunately, the rules precluded awarding such a prize posthumously.

At the time of Leavitt's paper on the period-luminosity relationship, there were only a couple of dozen Cepheids known, distributed across the sky. In 1914 Harlow Shapley followed up Leavitt's suggestion and used the 60-inch telescope of the Mount Wilson Observatory in California, at that time the largest telescope in the world, and found that the spectral lines were shifted blue-ward when the stars were brightest and red-ward when faintest. Radial oscillations meant the stars were pulsating, and he developed a theoretical explanation for this behaviour.

Since none of the Cepheids showed a parallax across the baseline of Earth's orbit of the Sun, measuring a distance by triangulation was not possible.[7]

Nevertheless, Ejnar Hertzsprung, at that time at Potsdam, reasoned that if the radial velocity of such a star was representative of its velocity perpendicular to our line of sight, then by measuring its proper motion it would be possible to calculate how far away the star would require to be for the presumed velocity on the plane of the sky to produce the proper motion measured. Although rather crude, this scheme would provide an initial approximation. Knowing the distance and the apparent magnitude of the star, it would be simple to calculate its true luminosity.

---

7. For telescopes of the early 20th century, parallax by triangulation was feasible only out to a distance of about 150 light years. As would later be determined, the nearest Cepheid, Polaris, the North Polar Star, is almost three times that distance. It is also a triple-star system, one member of which is the Cepheid with a period of 4 days.

In 1910 Lewis Boss of the Dudley Observatory in Albany, New York, had issued a catalogue of stellar proper motions. On consulting this, Hertzsprung found 13 stars that had light curves and periods similar to those in the SMC reported by Leavitt. On the basis of their proper motions he calculated the true luminosities of these stars, enabling him in 1913 to report using the Cepheids in the Milky Way to calibrate the period-luminosity relationship identified by Leavitt.[8] This was an excellent illustration of how the compilation of catalogues of stellar data by one observatory could enable an astronomer elsewhere to pursue research in a timely manner.

Although Henrietta Swan Leavitt received little popular recognition, her results directly facilitated the grand explorations of the Milky Way system by Harlow Shapley and, subsequently, of the realm of galaxies by Edwin Hubble. Hence it is no exaggeration to say her contribution was momentous.

Further Reading

Miss Leavitt's Stars: The Untold Story of the Woman Who Discovered How to Measure the Universe by George Johnson, W. W. Norton & Company, 2005.

The Glass Universe: The Hidden Story of the Women Who Took the Measure of the Stars by Dava Sobel, Fourth Estate, 2017.

---

8. 'About the spatial distribution of variables of the δ Cephei type', *Astronomische Nachrichten*, vol. 196, p 201, 1913.

Solar Observing

Peter Meadows

The Sun is a fascinating object for the amateur astronomer as it gives the opportunity to observe in detail the behaviour of our nearest star. Also, there is something different to see each day (99% of the time) and certain solar features can be observed to change in almost real-time. Observations can be made using minimal/modest equipment with contributions made to the Solar Sections of national and local astronomical organisations. It can also lead to the learning of new skills such as imaging or non-optical techniques. Finally, unlike observing other astronomical objects, there is no need to say up all night!

The Sun is at a distance of just 150 million kilometres from Earth and in comparison with other stars it is classified as a dwarf. The temperature at the core is around 15 million °C while on the visible surface of the Sun (the photosphere) it is about 5,500°C. Sunspots are cooler than the photosphere by about 1000°C and hence appear darker purely through comparison to the surrounding photosphere. Above the photosphere is the chromosphere, a layer a few thousand kilometres thick and which contains solar features that are far more dynamic to observe. Further out is the tenuous corona which is observed during total eclipses of the Sun has a temperature of 2 million °C. The Sun has a mean density of 1.41 times that of water, a mass of over 330,000 times that of the Earth and a volume equal to around 1.4 million times that of our planet. Aged at around 5,000 million years, the Sun is approximately halfway through its lifetime, and is composed primarily of hydrogen (~71%) and helium (~27%) with small amounts of oxygen, carbon, nitrogen and most other elements.

Warning: never look at the Sun with the naked eye or with any optical instrument unless you are familiar with safe solar observing methods.

White Light Observing

There are two main ways of observing the Sun in white light (i.e. in all optical wavelengths): by eyepiece projection or direct viewing.

The projection method is usually with a refracting telescope by using the telescope the 'wrong way round' in the sense of projecting an image of the Sun onto a piece of paper or card 30 cm or so away from the eyepiece end of the telescope. This approach avoids directly looking anywhere near the Sun in the sky. A projection box, such as shown in Figure 1 (left) can be easily constructed to fit between the eyepiece holder and the drawtube. This has the advantage of providing a sturdy holder for viewing the solar disk as well as casting a shadow to enable sunspots and faculae to be seen more easily. *Any finder telescope on the side of the main telescope must be covered to avoid any stray light reaching the observer.* Provided the telescope is on a sturdy mount (ideally this should be a driven mount) disk drawings can be used to record sunspot activity. For example, Figure 2 shows the largest sunspot group of the last solar cycle (the sunspot umbra and outline of the penumbra are shown). Reflecting telescopes can also be used to project an image of the Sun but it is recommended that the amount of light being received at the eyepiece be reduced by placing a mask over the telescope tube. In addition, plastic eyepieces must not be used, as they are likely to melt because of heat at the eyepiece end of the telescope.

The second approach is by direct viewing, which requires a suitable filter to be securely placed over the telescope objective lens before using the telescope

Figure 1. White light observing using eyepiece projection (left) and direct viewing (right). (Peter Meadows)

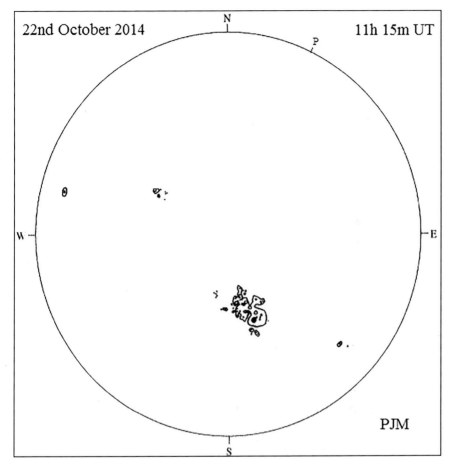

Figure 2. Example of a disk drawing. (Peter Meadows)

to view the Sun in the same way as any night time object. Figure 1 (right) shows a glass filter placed over the objective of a Meade ETX telescope – the filter has a tight fit over the telescope to ensure that it does not accidentally become loose. Note that the finder telescope is covered. Alternatively, solar safety film can be used either by using a purpose-made holder and film for your telescope aperture size or by constructing your own holder. In the latter case, the film should not be stretched over the holder as this can lead to pinholes being created. Irrespective of the holder type, the solar film should be checked regularly for pinholes and replaced if necessary (there are two types of solar film: one for visual observing and one just for imaging). Instead of using a filter over the main aperture, a Herschel Wedge can be placed at the eyepiece end of the telescope – this device significantly reduces the amount of light arriving at the eyepiece. Some observers prefer to view the photosphere using a Herschel Wedge, although they are likely to be more expensive than glass or solar film filters.

Whether observing by projection or direct viewing, the solar disk will usually show sunspots and faculae. The larger sunspots will show a dark umbra surrounded by a lighter penumbra. Sunspots also form into separate collections of sunspots or groups. By counting the number of sunspot groups observed on each day and averaging over a month (the Sun rotates once every 27 days as seen from Earth), the solar cycle of approximately 11 years can be monitored, such as shown in Figure 3, using an 80mm f/11 refractor and 6-inch (152mm) disk drawings. The red curve is a 13-month average that clearly shows the variation in solar activity over the last two solar cycles (and the double peaks for both cycles). It is usually straightforward to distinguish separate groups but this can be difficult during periods of high activity when careful examination of the solar disk is needed. Faculae are brighter regions that surround sunspot groups

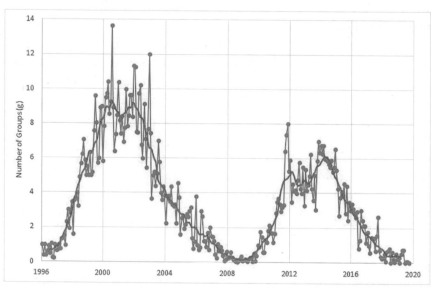

Figure 3. Monthly number of sunspot groups over two solar cycles. (Peter Meadows)

but they are only seen near the limb due to limb darkening. They usually appear before sunspots develop and after sunspots disappear.

Another way to monitoring the progression of the solar cycle is by using the protected naked eye. As with other optical observing, special care is required even with the naked eye. 'Eclipse Glasses' are ideal for this purpose as they are designed for observing the partial phase of a solar eclipse (note that they are not to be used with any other optical equipment). The same solar film as used for direct viewing can also be used. Using the protected naked eye will only enable the larger sunspots to be observed but during a solar cycle, many such sunspots can be seen. For example, during the most recently completed cycle, there were an estimated 50 separate naked eye sized sunspot groups. Once seen, they can be tracked across the solar disk from east to west. For particularly large sunspots, the shape of the sunspot can be seen (e.g. elongated). The difference in darkness can also occasionally be seen between the sunspot umbra and surrounding penumbra.

Narrow Band Observing

Hydrogen alpha is a deep-red spectral line of the Sun at a wavelength of 656.3nm and using a narrow band filter (typically 0.05nm) centred on this wavelength enables the more dynamic chromosphere to be observed. There are two types of hydrogen alpha equipment available to the amateur astronomer - dedicated hydrogen alpha telescopes or specialised hydrogen alpha filters added to night-time telescopes. The dedicated solar telescope was made popular and affordable by the introduction of the Coronado Personal Solar Telescope (PST) around 15 years ago. Now there are other manufacturers offering similar dedicated telescopes such as Lunt and Daystar. These telescopes are only used to observe the Sun. However, it is possible to purchase hydrogen alpha filters in conjunction with suitable energy rejection filters that can be fitted to ordinary refracting telescopes. These generally have larger apertures than the dedicated solar telescopes thus affording high resolutions viewing of the Sun at this wavelength. It is also possible to observe the Sun at other wavelengths such as the blue Calcium K spectral line but the features seen are far less dynamic and varied compared to hydrogen alpha. Most amateur observers concentrate on hydrogen alpha observing (also the eye becomes less sensitive to the blue part of the spectrum with age making it more difficult to observe the Sun visually in Calcium K).

The main features to observe in hydrogen alpha are prominences, filaments, plage and solar flares. Prominences are gaseous features that extend above the solar limb in a variety of shapes and sizes. They can change shape and brightness over timescales as short as a few minutes. In some cases material can be seen being ejected from the limb out into space. A particularly nice type of prominence is in the shape of a loop where movement within the loop can be seen over a period of say 30 minutes. Such events are exciting to watch especially when their physical size is appreciated (many tens of thousands of kilometres). If a prominence is observed on the eastern limb, it can sometimes be seen a day or so later over the solar disk in the form of a filament (the Sun rotates east to west). Filaments appear dark and occasionally the same physical feature can be seen as both a prominence and filament. Some filaments can extend over a significant portion of the disk and can change shape from day to day before, in the case of long duration filaments, appearing as a prominence once again on the western limb. Filaments can appear of different darkness's.

Plage are bright regions in the chromosphere of the Sun and are analogous to faculae in white light, appearing before/after the appearance of sunspot groups or even without the presence of sunspots (as is also the case for faculae). Unlike faculae, plage regions are not just seen near the limb but at any position across the disk.

Perhaps the most dynamic of all hydrogen alpha features are solar flares. Although rarely seen in white light, solar flares can be easily seen in hydrogen alpha within complex sunspot groups. When they occur they appear as bright regions amongst the sunspot umbra, filaments and plage of a group. They can appear suddenly, brightening rapidly before usually disappearing more slowly. So it is worth observing complex groups for extended periods of time to try and catch a flare. They can be of a variety of sizes and durations, and several flares can occur at the same time within the same complex group. Particularly large flares can, in some cases, lead to auroral displays appearing a few days later.

Imaging Solar Activity

Imaging solar activity can be as straight forward as taking a photograph of a screen at the far end of a projection box, such as shown in Figure 4, or attaching a DSLR camera to a telescope used for direct viewing of the Sun (i.e. with a

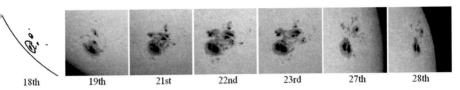

| 18th | 19th | 21st | 22nd | 23rd | 27th | 28th |

Figure 4. The passage of the largest group of Solar Cycle 24 in October 2014. (Peter Meadows)

full aperture filter as shown in Figure 1 [right]). In both cases, having a driven telescope mount will ease the taking of the photo (it is surprising how quickly the Sun appears to move through the field of view of a telescope).

Higher resolution images of the Sun in both white light and hydrogen alpha/Calcium K are usually taken with an astronomical webcam. This type of camera takes a series of short exposure images (frames) in the form of a movie AVI file for subsequent image processing. The advantage of taking many frames (several hundred or even thousands) is that frames with the best seeing (image steadiness) are used to create the final image (using a few tens up to a hundred or so selected frames). Free software is available to convert the AVI file to the final image: examples include RegiStax and AutoStakkert. Basically this software will register each frame to remove any movement between the frames in the AVI file (due to telescope tracking and seeing conditions), select and stack the best quality frames, apply a wavelet technique to improve the visual quality of the final image and finally provide image manipulation such as applying a gamma correction, cropping and so on. There are several manual steps for the user to apply but with a bit a trial and error and following tutorials available online, good quality images can be obtained.

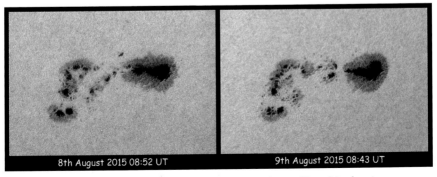

| 8th August 2015 08:52 UT | 9th August 2015 08:43 UT |

Figure 5. White light images of a sunspot group using a webcam. (Peter Meadows)

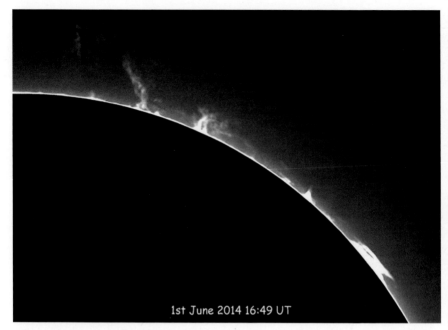

1st June 2014 16:49 UT

Figure 6. An active solar limb showing many prominences. (Peter Meadows)

An example of white light images is shown in Figure 5 using the Meade EXT 105 telescope shown in Figure 1 (right) and an Imaging Source DMK 31 monochrome webcam (many other suitable webcams are available). Here detail can be discerned within the sunspots as well as granulation within the surrounding photosphere. Subtle changes in the shape and size of sunspots

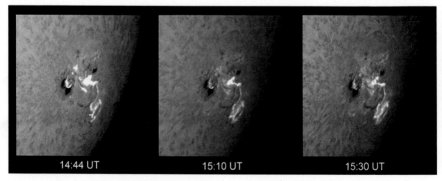

14:44 UT   15:10 UT   15:30 UT

Figure 7. The development of a large solar flare on 27 October 2014. (Peter Meadows)

over the period of a day can be seen. Figures 6 and 7 show prominence activity and the changes in the appearance of a large solar flare over the period of 45 minutes. These were obtained with a hydrogen alpha PST telescope, a Barlow lens (to enable focus to be achieved with the webcam) and the DMK webcam.

Submitting Observations

Although observing the Sun, or indeed any other astronomical object, by oneself is satisfying and of interest, more can be gained by combining observations with other observers. In the UK there are two national amateur astronomical organisations where your observations can be submitted on a monthly basis – the Society for Popular Astronomy and the British Astronomical Association. As these organisations have observers based in many locations around the world, gaps in observing due to the weather can be mitigated thus enabling the monitoring of solar activity on every day. It is also of interest to see what other observers are submitting in terms of observations and images. Both organisations show solar observations on their website and in their publications. Many local astronomical societies include active solar observers (it is also worth visiting your local society before purchasing any equipment for solar observing – societies will often have dedicated solar observing sessions that are open to all). Other countries often have similar national organisations (e.g. the American Association of Variable Star Observers) as well as numerous local astronomical societies.

Useful Links

AutoStakkert! – AS! Stacking Software **www.autostakkert.com**
British Astronomical Association **britastro.org**
Federation of Astronomical Societies **fedastro.org.uk**
RegiStax- Free image processing software **www.astronomie.be/registax**
Society for Popular Astronomy **www.popastro.com**

The Meteorite Age

Carolyn Kennett

Metals played such a crucial role in the advancement in human civilization that ancient eras, the Bronze and Iron Age, have been named after them. But pinpointing an actual date which humans first worked metal into objects is hard to do. In Europe this could have been as early as 7,000 BC. Brightly coloured copper and gold were utilised first, while extraction of tin and iron came later. A move from the Stone to the Bronze Age was made possible in about 2,500 BC due to the discovery that smelting copper with tin created the durable alloy bronze. The move to the Iron Age came much later on, about 1,200 BC. Iron is widespread in the geological landscape but extraction and smelting of this metal into a useable form proved incredibly complex. It is therefore perplexing to find iron objects which predate the Iron Age; some of these objects are at least 9,000 years old and are from as early as the first worked metals. However, another source of iron is non terrestrial. It arrives on Earth from outer space in the form of meteorites and it has been suggested that the utilisation of meteorites found on the surface could explain the presence of iron objects which predate the Iron Age. This would make meteorite objects some of the earliest encounters that humans had with metal.

It is not difficult to imagine that people would have admired and craved something unique, much in the way we do today. This is an easy concept to understand in a modern context, as in today's world 'things' matter; there is no denying we currently live in a material world. Objects enrich our lives; we use them as embellishments and allow them to express our identities and our hopes and fears. This is not a

A pendant from Umm el-Marra, Syria. Dating from 2,300 BC, this has been identified through the use of X Ray Spectroscopy as being made with meteoritic iron. (Albert Jambon)

modern phenomenon, and by walking around any museum we can see how longstanding the relationship to material objects is. In a world without modern materials and predating the times where the method to smelt iron was common knowledge, the properties of iron would have been special, and even without knowledge of its origin would have seemed otherworldly. Manipulation of this material most likely occurred with cold hammering as open fires would not have had the heat for smelting the metal. It would have shown the object to be shiny, durable, hard but malleable and made of something which could become useful such as a tool, sharp dagger, talisman or simple beads. There is no suggestion that in all cases people saw the meteorite fall and collected the deposit although in some examples there is a real possibility this was the case.

The examples of worked meteorite objects below will demonstrate how rare and highly prized this material was. This is not an exhaustive list as the exact number of prehistoric worked meteorite items changes continuously. This is not only due to new discoveries but also to re-examination of existing collections with better identification techniques and more precise dating of prehistoric objects. Historically the easiest and most accurate way to identify meteoritic iron has been by an inspection of its isotopes looking for raised nickel content within the object, in comparison to terrestrial iron which has no or little trace of nickel. Early methods were invasive and involved some destruction of the object to provide a sample. Further complicating identification is the fact that different nickel readings can occur depending on the location on the object tested, so one sample from the same object may result in a different reading to another.

Modern testing techniques aim for greater accuracy. This is achieved through the identification of the additional element cobalt, as well as a concentration of nickel in the specimen. Non-invasive techniques such as X-Ray Spectroscopy increasingly play a part as they do not involve damaging these often fragile objects. What the use of X-Ray Spectroscopy has revealed is that particularly corroded specimens may have little or no nickel content at all. These objects could have been wrongly identified as being terrestrial, but iron objects with no nickel content may still have been delivered from space. Albert Jambon, from the National Centre for Scientific Research (CNRS) in France, has imaged many of these fragile items with his X-Ray Spectroscope. He has identified a number of new specimens, explaining that "Some archaeologists were sceptical, as they thought that the amount of nickel found in Bronze Age iron tools was too low to consider them of meteoritic origin. But I'm a trained cosmochemist, so I knew the problem was just corrosion. And I was able to show that nickel was leached away during corrosion." Modern techniques are increasing the number of known objects which are meteoritic in origin.

Only sixteen known iron objects predate 3,000 BC and these are outlined in the table below. The earliest worked meteorite objects, three beads, were discovered in 2014 at Lake Świdwie, Poland. This exciting discovery was

Albert Jambon testing a meteorite object with his X-Ray Spectroscope. (B. Devouard)

mentioned in the *Yearbook of Astronomy 2020*. Dating from 7,000 BC this is not only the earliest example of a worked piece of meteoritic iron, but one of the earliest worked metal objects in the world. Due to the location of its discovery the significance cannot be overstated; all the other objects from this period come from the cradle of civilisation and northern Poland sits well outside this region. Prior to this the nine blackened beads found in a pre-dynastic cemetery near el-Gerzeh, Egypt were the oldest known worked iron meteorite objects. The beads were scanned and revealed the distinctive Widmanstätten structure found in iron meteorites. During this investigation of the beads Professor Thilo Rehren of the Petrie Museum (University College London) discovered a little about the technique used to create the objects saying "The shape of the beads was obtained by smiting and rolling, most likely involving multiple cycles of hammering, and not by the traditional stone-working techniques such as carving or drilling which were used for the other beads found in the same tomb." Furthermore he felt that the Egyptians had an advanced understanding of the material they were using, suggesting that they had worked with it before. The final object in the table below is a bit of an anomaly itself. Identified as terrestrial smelted iron, this four-sided tool was discovered in a tomb in Samarra, Mesopotamia. One suggestion is that the iron used in this object had been obtained as a by-product from extraction of another metal, although that does not explain how the society overcame the difficulties achieving the high temperatures needed for the smelting process. It is a real possibility that retesting using Albert Jambon's spectroscopy technique would identify it as being meteoritic in origin.

OBJECT	DATE	LOCATION OF FIND	OTHER OBSERVATIONS
Beads x3	7,000 BC	Lake Świdwie, Poland	Worked – found in Shamans Hut
Balls x3	4,600–4,100 BC	Tepe Sialk, Iran	Polished but unworked – hard and heavy – found in palace.
Beads x9	3,200 BC	El-Gerzeh, Egypt	Widmanstätten structure . Grave goods. 9% nickel
Four-sided Tool	5,000 BC	Samarra, Mesopotamia	First non-meteorite smelted object? Grave goods.

During the following millennium the number of iron objects increase, although they are still rare in comparison to other metallic objects made from copper or gold. Iron was still highly prized and treated as a precious metal. All the

The Royal Tomb at Alaca Höyük, Turkey where the hoard of ceremonial meteorite objects dating from 2,400 BC were discovered. (Bernard Gagnon)

discoveries are objects of significance and include jewellery, decorative items and ceremonial weapons and have been discovered in sealed hoards, near temples or deliberately buried in rich graves. Rather than being objects intended for everyday use, they were luxury goods, much sought after by the populace. There are nine confirmed meteorite objects from this period, including the hoard from Alaca Höyük.

One unanalysed object from this period is an iron sword found in a royal tomb in Dorak, Egypt dated to 2,400 BC. A beautiful early example of a ceremonial weapon, it has an obsidian holt carved into two leopards, inlaid with gold and amber spots. Unfortunately this example was excavated surreptitiously and is now lost, our knowledge of this object coming from a cartouche of the Egyptian pharaoh Sahure, second ruler of the Fifth Dynasty. This illustrates the problem in putting together a comprehensive list. There are at least another ten examples which have not been confirmed as being meteoritic in origin and which therefore do not make it onto the following list.

OBJECT	DATE	LOCATION OF FIND	OTHER OBSERVATIONS
Fragment	3,100 BC	Urak, Mesopotamia	Found in temple
Disc	2,500 BC	Ur, Mesopotamia	10.9% nickel. Found in tomb.
Pendant	2,400 BC	Umm el-Marra, Syria	Found in tomb
Pins x2	2,400 BC	Anatolia, Turkey	Found in tomb
Plaque	2,400 BC	Alaca Höyük Turkey	Found in tomb
Dagger	2,400 BC	Alaca Höyük Turkey	First discounted as terrestrial iron but retested and identified as meteoritic in origin.
Mace Head	2,400 BC	Alaca Höyük Turkey	Found in tomb
Amulet	2,100 BC	Deir el Bihari, Egypt	Tomb of Princess Aa Shait Dynasty XI

The late Bronze Age (2,000–1,200 BC) has produced a greater number of confirmed meteorite examples. Twenty three have been identified to date with over fifty objects still to be analyzed. What is interesting is that no smelted objects from this period have been identified. With the Iron Age about to burgeon we would almost expect some isolated examples of terrestrial smelted iron objects, so the lack of them is curious in itself. The meteorite objects found are geographically scattered, albeit with nineteen of them coming from

Tutankhamun's tomb in Egypt. Surprisingly these objects came from three different meteorites rather than from a single iron meteorite fall. This suggests that the meteorites were being actively sought by the populace to be made into grave goods. Within the list are objects which are more utilitarian, such as the chisels found in Tutankhamun's tomb. These were still highly prized grave goods and do not demonstrate a regular usage of the material.

OBJECT	DATE	LOCATION OF FIND	OTHER OBSERVATIONS
Fragment	1,600 BC	Crete	20lb piece unworked found in Minoan Palace
Axe	1,400 BC	Ugarit, Syria	Decorated with gold – ceremonial
Axes x2	1,400 BC	China	Shang Dynasty
Dagger	1,350 BC	Egypt	Tutankhamun Tomb
Headrest	1,350 BC	Egypt	Tutankhamun Tomb
Bracelet	1,350 BC	Egypt	Tutankhamun Tomb – Eye of Horus
Chisels x16	1,350 BC	Egypt	Tutankhamun Tomb – found in box together

Stony meteorite falls are far more prevalent than iron ones, their similarities to terrestrial rocks and the fact that many break up into smaller pieces would make the utilisation of large iron meteorites a more regular occurrence. That does not mean that stony meteorites have not been collected by humans in the past. One example was found during an archaeological dig in 1989. Discovered in a pit at Danebury Hill Fort and identified as a piece of H5 ordinary chondrite the meteorite was dated to 2,350 ± 120 year BP (Before Present). Unusually it was found in an unworn, fresh condition and has a weathering index of W1/2. The conclusion is that it had either fallen directly into the manmade pit just before it was filled or had been found and placed in the pit deliberately. During this period Danebury Fort was heavily occupied and a deliberate placing suggests that the

The Danebury Meteorite – a stony meteorite discovered in a pit at Danebury Hill Fort, Hampshire, England during an archaeological dig in 1974 was identified as being a meteorite during 1989. (The Open University)

meteorite fall may have been witnessed and that the object had been revered by the owner before offering it as a gift by placing it in the pit.

It is worth considering if communities understood the relationship between the object and its otherworldly origin. Evidence from the names given to the material suggests they did. The Hittites (from Anatolia, modern-day Turkey) called iron *An-Bar Ge*, *nepisai* or 'black iron of the heavens', while the Egyptian term *bia' n pet* means 'iron of heavens' and both suggest an understanding of the relationship. The Egyptian term came from around the time of the 18th Dynasty, around 1,300 BC. It was a new description at that time and linguists believe it relates to an observed fall. The Gebel Kamil impact crater in the East Uweinat Desert in south western Egypt was due to an iron meteorite which fell around 5,000 years ago and could be the site of this observation, but without written witness accounts it is hard to pinpoint the actual event.

Many stories of impacts have been lost to prehistory; therefore it is worth exploring a modern example about an observed fall and the resulting human perceptions of the meteorite. In Duruma, Kenya in East Africa a community collected a one pound meteorite which fell on 6 March 1853. Local German missionaries tried to buy this from the Wanika tribe but they refused to sell it and started to worship it as a god. They built a temple to enclose it, annotated it with oil and pearls and even clothed it. For three years they worshiped this newfound deity, until the Masai attacked the Wanika village, burning it to the ground and killing many. They decided that the object was a poor protector and gladly gave it away. The meteorite is now in the Academy of Sciences of Munich, Germany and is a good illustration of how fickle humans can be especially when superstition rules. Another example is that of the Hopewell Meteorite from Hopewell Mound, Ross County, Ohio. This meteorite was fashioned into beads which were found in ritual settings, such as burial mounds. The reverence paid to these objects suggests that the people may have understood its cosmic origin; they had also collected iron meteorites from Brenham, Kansas, which were made into more mundane objects such as axes, chisels and drills.

This is not the only instance of meteorites being used as mundane objects in Kansas. Meteorite objects were utilised by the first settlers of the Kiowa area for their base properties and were put to use in the most ordinary ways; such as anvils and nut crackers and weights to hold down rain barrel covers and stable roofs. This everyday use displayed a complete lack of understanding of the origin and rarity of the material. Furthermore, the objects were often considered a nuisance as they would break the settlers' valuable ploughing machines.

Overwhelmingly meteorite finds have been prized and that is demonstrated in the type of objects they were fashioned into, or by the location in which they were discovered. Although rare, there is no doubt that a number of early meteorite worked objects are still lying in collections around the world waiting to be identified. Less invasive techniques are revolutionising the way these objects can be identified, although in some ways is a race against time as unfortunately many of them are in a fragile state and rusting away. To understand humanities relationship to meteoritic iron objects and early metallurgy it is critical that these objects are identified and classified in the correct manner. It would be useful to have a special meteorite category at museums for manmade meteorite artefacts; after all, we could have been living through an Age of the Meteorite.

Following their examination in 1971 of ancient Chinese axe blades at the Freer Gallery of Art in Washington, D.C., investigators Roy S. Clarke, Jr. (Associate Curator, Mineral Sciences, Meteorites), William Thomas Chase, and Rutherford John Gettens were able to determine that two ancient Chinese axes in the Freer collection were made from iron meteorites. (Smithsonian Institution)

'A Dignity That Insures Their Perpetuation' Obsolete Constellations and the Making of the Modern Night Sky

John C. Barentine

Joel Dorman Steele, a prolific American textbook author of the late nineteenth century, noted that the Western constellation figures bear little resemblance to the mythical heroes, exotic beasts, and apparatus of the arts and sciences they represent. "The likeness is purely fanciful," Steele wrote in 1899. "Not only are the figures uncouth, and the origin often frivolous, but the boundaries are not distinct. Stars occur under different names; while one constellation encroaches upon another." But Steele argued that deference to tradition and the passing of many centuries since their creation insulated the constellations from further alteration. "However the constellations are thus rude and imperfect," Steele argued, "there seems little hope of any change. Age gives them a dignity that insures their perpetuation."

A canon of 88 constellations, established by the world community of professional astronomers, composes the entire contents of every modern star map. This number is as arbitrary as the construction of constellations itself. The lore of the night sky, at least in the Western tradition, is taken as the received wisdom of the ages, and by rights it is. Some of the figures constituting this set may have been first identified by humans over 20,000 years ago, while the last few added to the collection made their debut in the mid-eighteenth century. The process by which some were kept while others were discarded did not take place in a vacuum, and it owes a great debt to the two millennia that preceded the selection.

The Western view of the night sky seen as seen in the northern hemisphere was largely complete by the second century AD when it was committed to history in the pages of a text that has come to be known as the Almagest. The treatise was written by Claudius Ptolemy, an ethnic Greek or Hellenized Egyptian living in Alexandria who may have held Roman citizenship. Among its thirteen 'books' on the apparent motions of the planets, an outline of Aristotelian cosmology, and eclipses of the Moon and Sun is found a catalogue of 1,022 stars divided into 48 constellations. In addition to the 12 figures of the zodiac, 36 other constellations define the night sky as it was known to the Greeks, who never ventured further south than the Nubian frontier of southern Egypt near latitude 24° north. That left nearly one-third of the night sky unknown to them. (Figure 1)

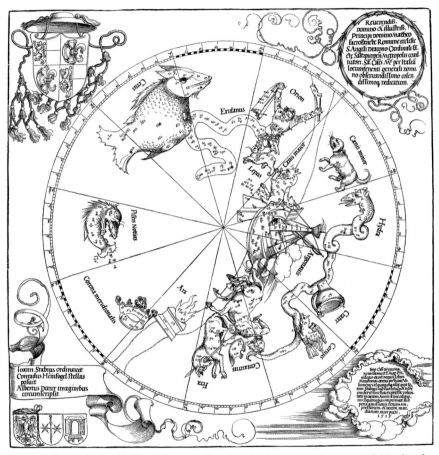

Figure 1. The southern night sky as shown in Albrecht Dürer's 1515 map *Imagines Coeli Meridionales*. The map, centred on the south celestial pole, shows only the constellations identified by Ptolemy in the second century AD. The conspicuous empty circular region at centre-left indicates the part of the night sky permanently below Ptolemy's southern horizon in his time; the gap between the centre of this circle and the pole on Dürer's map indicates precession of the equinoxes by about eighteen degrees during the intervening fourteen centuries. (Wikimedia Commons/Albrecht Dürer/Imagines Coeli Meridionales)

Ptolemy's view prevailed at the end of classical antiquity and persisted in Western thought for more than a thousand years. However, Ptolemy merely cribbed from the second century BC Greek astronomer Hipparchus. The Greeks in turn learned the same constellations in their interactions with the civilizations of the ancient Near East. There is evidence that the core set of Ptolemaic figures was practically 'baked in' by the end of the twelfth century BC, its iconography firmly associated with native Greek myths.

Even the Christianization of Europe that began in late antiquity did not displace the pagan myths from their place in the cosmos; as late as the early seventeenth century, some ambitious celestial cartographers tried unsuccessfully to re-brand the Greek figures with themes of the Old and New Testaments. (Figure 2) Other mapmakers, respecting the Ptolemaic convention, noted astutely that Ptolemy himself referred to certain stars as *amorphotoi*

Figure 2. The now-obsolete Ptolemaic constellation Argo Navis became Noah's Ark in the Christianized firmament of Julius Schiller's *Coelum Stellatum Christianum* (1627). (Wikimedia Commons/Julius Schiller)

(meaning "unformed") and not belonging to any of his figures. They sensed and seized upon an opportunity for fame and profit by claiming these stars for their own designs.

Like the sky above, the Earth below began to open up to Europeans in the fifteenth and sixteenth centuries. The discovery of the lands of the western hemisphere around 1500 validated ideas about a large but navigable world whose continents were surrounded by contiguous oceans. An age of exploration saw the "new" world colonized by the old as Europeans ventured for the first time to the planet's southernmost reaches. There they encountered an utterly unfamiliar sky full of stars that were unknown to the Greeks.

New southern constellations were first identified as a navigational aid that could be communicated to other explorers through maps. Some of the figures were literal representations of navigational instruments: Circinus (the Compass Circles), Horologium (the Pendulum Clock), Octans (the Octant). Others referred to strange new animals the Europeans encountered on their voyages: Tucana (the Toucan), Chamaeleon (the Chamaeleon), Apus (the Bird

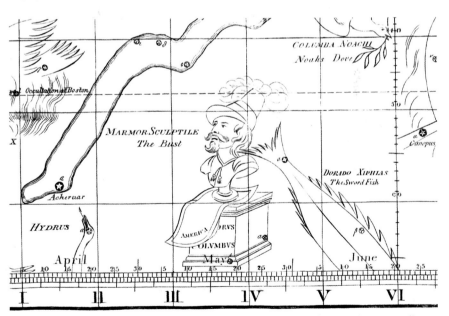

Figure 3. In 1810, William Croswell took stars largely comprising the modern constellation Reticulum (the Reticle) to form Marmor Sculptile, a marble bust depicting the Italian explorer Christopher Columbus. (U.S. Library of Congress / William Croswell)

of Paradise). While these figures are counted among the modern constellations, others were proposed and enjoyed brief popularity before falling into disuse such as Rhombus (the Bullroarer) and Polophylax (the Guardian of the Pole). The traditional 'discoverer of the New World' was even honoured with his own figure, Marmor Sculptile (the Bust of Christopher Columbus), by the American cartographer William Croswell on his 1810 *Mercator map of the starry heavens*.

As the initial phase of global exploration came to a close in the eighteenth century, the makers of star charts dealt with Ptolemy's *amorphotoi* by daring to create their own constellations, inserted strategically between Ptolemaic figures. They staked their claims on the maps they drew, published and distributed in a kind of popularity contest: new constellations garnering the greatest public acclaim would make their way onto others' charts and slowly assume a place in a growing de facto canon. Although it has ancient origins as an asterism, Coma Berenices (Berenice's Hair) was the first new northern constellation introduced since antiquity that is still found on modern charts, appearing on a celestial globe in 1536. Other figures, including Monoceros (the Unicorn) and Camelopardalis (the Giraffe), appeared later in the sixteenth century. The only manifestly Judeo-Christian symbol still found on modern star maps was introduced in this era: Columba Noachi (later just "Columba") representing the dove Noah released from the ark in order to learn whether the global flood was subsiding.

The collection of constellations defining the northern night sky as we know it today was essentially complete at the end of the seventeenth century. The latest entrants were submitted by the Polish astronomer Jan Heweliusz, publishing under the Latinized 'Johannes Hevelius', in *Firmamentum Sobiescianum* (1690); these finishing touches included Lynx (the Lynx), Lacerta (the Lizard), and Canes Venatici (the Hunting Dogs). Although these stars remained in dispute through the end of the eighteenth century (Figure 4), no constellation suggested after Heweliusz survived the twentieth-century cut that defined the modern canon.

The photographic process revolutionized astronomy when it was introduced in the latter half of the nineteenth century, beginning a process by which visual observing would be rendered all but obsolete in terms of value to the scientific method. Among its other benefits, photography enabled the precise measurements of the positions of stars, obviating the need for the aesthetically pleasing, hand-drawn sky charts of the past. Around the same time, the

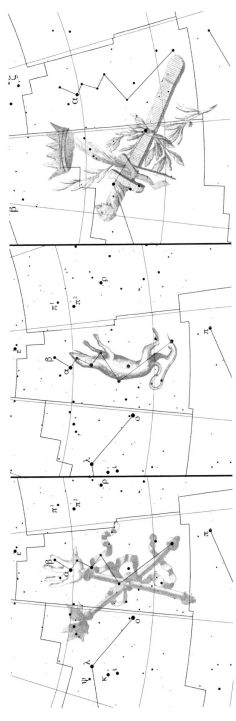

Figure 4. Three attempts in historic star charts to map faint 'unformed' stars between Pegasus, Andromeda, and Cepheus. In 1679, Augustin Royer created Sceptrum et Manus Iustitiae (left), depicting the royal regalia of France. Eight years later Jan Heweliusz formed the modern constellation Lacerta, the Lizard, in *Firmamentum Sobiescianum* (1690, centre) using many of the same stars. Almost a century after its publication, Johann Elert Bode appropriated some stars from both Lacerta and neighbouring Andromeda to form Honores Frederici, a constellation honouring the achievements of the recently deceased Frederick II of Prussia. In each panel, modern constellation boundaries are indicated along with some lines of constant right ascension and declination. (Left: author photo / Corbinianus Thomas; centre: Wikimedia Commons / Johannes Hevelius; right: Wikimedia Commons / Johann Elert Bode)

burgeoning field of variable star studies -- still largely carried out through the diligent efforts of visual observers -- brought about a set of conditions that demanded a more precise definition of where one constellation formally ended and another began. The need to clearly delineate such boundaries resulted in the declaration of the particular set of constellations, and their limits, found on every modern star chart.

As photography made the process of collecting astronomical data more objective and quantitative, the observation of variable stars was coming into vogue. Since by tradition variables are labelled according to the constellation in which they appear, clear boundaries between constellations became an issue of serious concern such that astronomers properly communicated among themselves to which star one observation or another pertained. But because boundaries varied from chart to chart based on the preferences of the cartographers who drew them, confusion reigned. Furthermore, the precession of equinoxes, resulting from the slow wandering of the Earth's rotational axis in space, could cause a particular variable star to apparently cross anyone's arbitrary dividing line over time. The problem remained unsolved until the decade following the First World War, when the professional astronomical community imposed a sense of order on the night sky meant to be final.

The newly established International Astronomical Union (IAU) met for the first time at Rome, Italy, in 1922. Among the topics the congress addressed was the architecture of the night sky itself. At its first General Assembly, IAU members settled on a list of 88 'modern' constellations. In the process, they discarded roughly 25 figures that appeared on multiple charts between the seventeenth and twentieth centuries. We do not know many of the details of the process by which the decisions were made. All 48 of the Ptolemaic constellations were kept, plus contributions from mapmakers and astronomers to the end of the seventeenth century. Clear boundaries were established, following lines of constant right ascension and declination tied to the epoch 1875.0, which could be adjusted in the future so as to prevent precession from changing the identification of stars with particular constellations. By 1930, when the final set of charts was printed, the case for further changes was essentially closed. From then on, constellations were defined regions of the night sky, contiguous with one another in every instance, and whose demarcation was tied to the system of equatorial coordinates that remains widely used.

The nearly three dozen constellations cut from lists of prospective inclusions in 1922 constitute a collection of figures as diverse as the 88 constellations that the IAU dubbed official. Just like the assortment of suggested additions adopted during the time between Ptolemy and Rome, the discarded figures were born of many of the same motivations: flattering patrons, commemorating nationalist symbols, honouring significant individuals, and recognizing the contributions of science and technology. The cultural attitudes of post-WWI Europe, for example, may have influenced dropping constellations honouring technological advances attributed to ethnically German people, such as the Electric Generator (Machina Electrica) and the Printing Press (Officina Typographica). The IAU similarly turned down two constellations introduced in the late eighteenth century to recognize the discoveries of the Anglo-German astronomer William Herschel (Telescopium Herschelii Major and Minor). But it also declined to canonize Custos Messium (the Harvest Keeper), a somewhat less pointed reference to the eighteenth century French comet hunter Charles Messier.

Some formerly independent constellations lost their status and were relegated to asterisms, or small figures consisting of prominent stars but which are not recognized as constellations in their own right. These range from Caput Medusae (the Head of Medusa; Figure 5), carried in the right hand of Perseus, to Lochium Funis (the Log and Line), a figure that once ranked independently alongside other pieces of the Ptolemaic figure Argo Navis, the Argonauts' ship from classical mythology. The latter is an example of a cartographer, in this case the French Abbé Nicolas Louis de La Caille, who in 1752 dispensed with the large and unwieldy Argo by breaking it into pieces and giving the components identities as their own constellations: Carina (the Keel), Puppis (the Poop Deck) and Vela (the Sails). These remain canonical, but Lochium Funis was fully absorbed into the stars of the neighbouring Pyxis (the Mariner's Compass) by the end of the nineteenth century.

In an era during which there was no consensus among astronomers about the exact composition of the constellation canon, some enterprising astronomers formed new figures in order to flatter their benefactors. At least nine such figures once appeared on European star charts, including Gladii Electorales Saxonici (the Crossed Swords of the Saxony Electorate, honouring Johann Georg III, the Holy Roman Elector of Saxony); Honores Frederici (Frederick's Glory, honouring Frederick II of Prussia, seen here in Figure 4); Sceptrum et

Figure 5. Some figures now considered asterisms had former lives as constellations unto themselves, such as Caput Medusae (the Head of Medusa), shown here in Corbinianus Thomas' *Mercurii philosophici firmamentum firmianum* (1730). (Author photo/Corbinianus Thomas)

Manus Iustitiae (the Scepter and Hand of Justice, honouring Louis XIV of France, also in Figure 4); and Taurus Poniatovii (Poniatowski's Bull, honouring Stanisław II Poniatowski of Poland).

A more brazen approach involved repurposing ancient figures for decidedly contemporary reasons. Corbinianus Thomas, a Benedictine monk who lived in and around Salzburg, Austria, in the eighteenth century, tried unsuccessfully to transform the Ptolemaic figure Corona Borealis (the Northern Crown) into Corona Firmiana, commemorating the presumed achievements of Leopold

Anton von Firmian, Prince-Archbishop of Salzburg. But it was Jan Heweliusz who gave our star charts the only instance in which a beneficiary successfully acknowledged his patron's largesse: Scutum, originally called 'Scutum Sobiescianum' (Sobieski's Shield), which recognized the material support of John III Sobieski, King of Poland and Grand Duke of Lithuania.

The 'lost constellations' became so for a variety of reasons. Chief among these is that they were formed from relatively faint and inconspicuous stars that did not lend themselves well to readily recognizable patterns. Those that were not easily sighted tended to disappear from star charts first for lack of

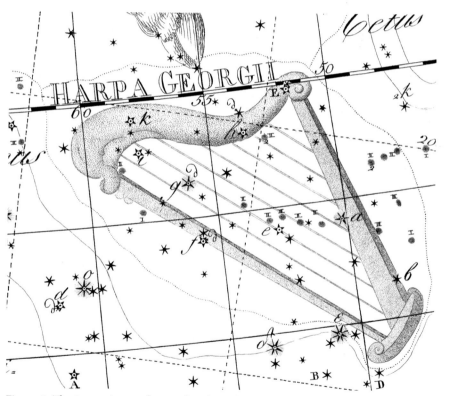

Figure 6. The inconspicuous figure of Psalterium Georgianum, shown labelled as "Harpa Georgii" on Plate XII of Johann Elert Bode's *Uranographia* (1801). The example also illustrates the arbitrary nature of constellation boundaries before the International Astronomical Union's uniform definitions were published in 1930: note that Bode carefully drew the dotted line separating this constellation from neighbouring Eridanus in order to exclude from the former the bright star ε Eridani (lower right). (Wikimedia Commons/Johann Bode)

prominence. An example is shown in Figure 6: a collection of stars west of Orion and south of Taurus once called Psalterium Georgianum, the figure of a harp whose formation in 1789 was intended to honour King George III of the United Kingdom. That its brightest star was the fourth-magnitude 10 Tauri did not endear this tribute to astronomers, and it vanished from contemporary charts before the end of the nineteenth century.

In several cases, the increasingly contrived figures were identified with subjects and symbols too wrapped up in the history and culture of one particular (European) nation to enjoy broad appeal across borders. More fundamentally, it seems that a sense developed in the about the nineteenth century that humanity simply did not need figures to represent every single group of stars visible to the naked eye.

The story of the obsolete constellations may otherwise be considered a curiosity of the astronomy history collector's cabinet, but they have significance even in the modern world. Their story may be a historical relic, but they did not actually disappear from the night sky when they vanished from star charts. In the same way that nations of old are still readily identifiable among the demographic distributions of people on Earth even as the distinctly human creations of political maps and artificial borders willed them away, the 'lost' constellations may yet be found in tonight's sky if one simply knows where to look.

Their proliferation corresponds to an 'age of discovery' when European explorers vastly expanded the boundaries of the world known to Western civilization and science and technology grew by leaps and bounds. They document a time near the end of the continuum between antiquity and the modern world, which saw the emergence of astronomy (and later, astrophysics) as a fully empirical science. This may have diminished the importance of constellations generally, but they remain signposts for people just discovering astronomy, many of them amateurs for whom the constellations are their guides around the night sky. And, lastly, the obsolete constellations beg the question of what the figures we place in the night sky will mean to people of the future: of today's 'official' constellations, which will future humans choose to keep and which will they discard?

Lunar Volcanism
The View 50 Years after Apollo

Lionel Wilson

In the decades before the Apollo missions, some scientists thought that many of craters on the Moon were volcanoes of some kind. However, as we recognised impact craters on Earth, and understood more about them, we realised that almost all of the lunar craters were produced explosively, by high-speed impacts of asteroids and meteoroids. Then we started to get images of the lunar surface from the five Lunar Orbiter spacecraft used to select landing sites for the Apollo missions and began to see features that looked very like large lava flows. The ones in the interior of the giant impact basin Mare Imbrium were particularly well photographed by the Apollo astronauts while in lunar orbit (Figure 1).

Planets have volcanic activity because internal heat sources cause some of the mineral grains forming rocks to melt. The main long-term planetary heat source is the decay of naturally radioactive elements. A planet's heat production is proportional to the amount of the heat source that it contains, which in turn

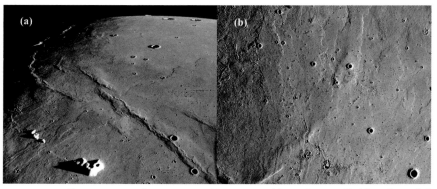

Figure 1. Lava flows in Mare Imbrium. (a) An oblique view of flows that travelled from lower left to middle right of the image. Part of NASA Apollo image AS15-M-1555. The ridges aligned upper left to lower right formed after the flows were emplaced. (b) NASA Apollo image AS15-M-1702 shows a vertical view of flows that travelled from the bottom to the top of the image. The large numbers of impact craters indicate the great age of the flows.

is proportional to the volume of the planet, i.e., to the diameter cubed. A planet only loses heat through its surface, so the rate of heat loss is proportional to the surface area, i.e., to the diameter squared. The Earth is about 4 times the diameter of the Moon, so it produces heat $4^3 = 64$ times faster than the Moon and loses heat $4^2 = 16$ times faster than the Moon, making it $64/16 = 4$ times better at conserving its internal heat. As a result, the Earth is still vigorously volcanically active but, as confirmed by dating samples returned by the Apollo missions, the Moon has probably not had an eruption for 1000–1500 Ma, roughly the last 20% of its life.

Almost all the Moon's volcanic rocks are located inside, or very near, the giant impact basins, which are themselves simply very large impact craters. Furthermore, virtually all the basins on the side of the Moon that always faces the Earth contain volcanic rocks, whereas far fewer of those on the far side do so. This difference is due to the internal structure of the Moon. The Moon's shape has been measured from images, and its internal structure has been probed by looking at the effects of lunar gravity on spacecraft orbits, especially those of the twin GRAIL satellites. It has become clear that the crust of the Moon, a layer of low-density rocks called anorthosites formed very soon after the Moon accreted, is about 30 kilometres thick on the Earth-facing side of the Moon and about 50 kilometres thick on the far side. Beneath this low-density crust the bulk of the Moon consists of its mantle, composed of denser rocks. Recent research has shown that many of the lavas erupted onto the surface started life when rocks deep within the mantle (as much as 500 kilometres beneath the surface, nearly one third of the way to the centre of the Moon) started to melt.

The liquid rock, called magma, was less dense than the unmelted mineral grains around it and seeped slowly upwards, eventually creating cracks, called dykes, that slowly filled with magma. The dykes grew larger and more buoyant, and eventually broke away from their source regions to rise very quickly toward the surface. When they reached the surface, these dykes poured out magma from fissure vents typically 20 kilometres long at a very great rate, up to one million cubic metres per second, more than one thousand times faster than typical eruptions seen today on Earth in places like Hawai'i and Iceland. Although the magma was very buoyant in the mantle, it was denser than the crust, so each dyke slowed down as less and less of it remained in the mantle, reducing the eruption rate at the surface. Eventually a density balance was reached, with part of the dyke in the crust and the rest still in the mantle. Meanwhile the crustal rocks on either side of the dyke, which had been pushed apart by its arrival, started to relax, and eventually squeezed most of the rest of the magma out of the dyke at a slow rate. This pattern of high initial eruption rate followed by a steady decline is common in eruptions on Earth.

An early finding from the chemical analysis of the Apollo lunar samples was that the Moon was depleted in many volatile elements and compounds. This was presumably the result of the unusually rapid and very hot accumulation of the Moon from debris thrown out by a giant impact on the early Earth. At first, we thought that the Moon contained essentially no water, the commonest gas released in volcanic eruptions on Earth, and only very small amounts of other gases. It was realised that the main gas involved in lunar eruptions was carbon monoxide, produced in a way unfamiliar on Earth, by a chemical reaction between metal oxides and graphite (i.e. carbon). As time has passed and scientific instruments have become ever more sensitive, re-analyses of the returned Apollo rock samples have shown that the Moon's mantle contains a wider range of volatile compounds than we originally thought, including small amounts of water, sulphur compounds, fluorine and chlorine.

During eruptions on the Moon these volatiles were released, along with carbon monoxide, as gas bubbles when rising magmas reached the low pressures near the surface. The Moon has probably never had a significant atmosphere, so with no pressure to constrain them, all of the gas bubbles in the magmas did their best to expand to an infinite volume as they reached the surface. In doing so, they tore the magmatic liquid apart into enormous numbers of tiny liquid droplets, typically very much less than a millimetre in size, and blasted these out in vigorous explosive eruptions, the equivalent of what we call fire-fountain or lava-fountain eruptions on Earth. In eruptions with a relatively high gas content and low magma discharge rate, the droplets cooled as they travelled through the vacuum and were solid when they reached the ground, forming the tiny glassy spheres found in many of the Apollo samples (Figure 2). In other cases, with a high eruption rate and small gas content, droplets were so crowded together while in flight that most of them could not radiate heat out into space and so landed back on the ground still liquid, just as hot as when

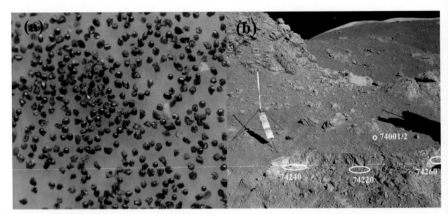

Figure 2. Lunar pyroclastic glass spheres. (a) Laboratory image of sample 74220 of the orange glass droplets found by Apollo 17 astronaut Harrison Schmitt. Droplet size range 90–150 microns. NASA Apollo image S73-15085. (b) The site where the orange glass droplets were found. NASA Apollo image AS17-137-20990.

they left the vent. In that case they coalesced as they landed into a lava lake, which overflowed to feed a lava flow. Figure 3a shows this happening in a lava-fountain eruption on Earth, and Figure 3b attempts to show how the lunar version might have appeared.

The combination of very high eruption rates and very hot lava on the Moon sometimes caused a process not commonly seen on Earth. Here, the flow of lava is generally quite smooth (the technical term is laminar), and a layer of chilled lava builds up at the bottom and top of the flow, so heat loss to the ground and to the atmosphere is not very great. On the Moon, the lava moved in a turbulent manner, meaning that hot lava from the interior of the flow

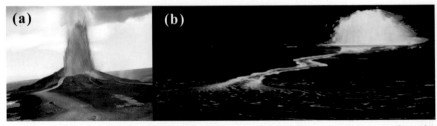

Figure 3. Typical lava fountains on the Earth and Moon. (a) A 180 metre high fountain from the Pu'u 'O'o vent on Kilauea volcano, Hawai'i. USGS image. (b) Simulation of how a typical lava fountain would have appeared on the Moon at night. The scale is the same for both images.

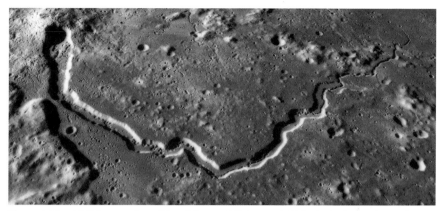

Figure 4. The 115 kilometre long Rima Schroeteri (Schroeter's Valley), the largest lunar sinuous rille. Part of NASA Apollo 15 image S71-44666.

was being constantly moved to the bottom of the flow where it heated up the ground beneath. This eventually caused the ground to start to melt, and the molten surface rock was incorporated into the flow, excavating a depression. As time went on, the lava flow subsided into the floor of the depression that it had created for itself. The resulting channel is called a sinuous rille (Figure 4). Of course, at the same time that the turbulent lava was heating up the ground beneath it, it was radiating heat into space, so the entire flow cooled more quickly than it would have done if it had been laminar. Eventually it ceased to be able to erode the ground beneath it and flowed on as a normal lava flow on the surface until it eventually cooled to the point where it had to stop moving. The lava in the lake around the vent that fed the flow was also turbulent, and so it eroded the floor of the lake to form a depression around the vent, as shown in Figure 4. This is enormously useful, because the radius of the depression shows how far the molten droplets leaving the vent travelled, and hence their launch speed, which is in turn determined by how much gas was released by the magma. With this estimate of the gas content, we can use the fact that the lake lava was turbulent to get a minimum estimate of the eruption rate – the volume of lava released per second. These numbers are our most direct evidence for the assertion made earlier that many of the ancient eruptions on the Moon were much more vigorous than any that we see on Earth today.

The most recent discoveries about volcanism on the Moon have come from the extremely successful Lunar Reconnaissance Orbiter mission (LRO), which is still on-going and can image features on the lunar surface smaller than a metre. Two new types of volcanic feature have been recognised: Irregular Mare Patches (IMPs – Figure 5) and Ring Moat Dome Structures (RMDSs – Figure 6). There is still considerable debate about the exact origins of these features, though their association with volcanic rocks is totally clear. The IMPs are generally a combination of depressions containing material brighter than its surroundings between rather bulbous lobes of darker material (Figure 5). Some are associated with what appear to be vents for lava flows (Figure 5a) and other, smaller ones, just appear in patches (Figure 5b). The association between the depressions and bulbous parts suggests that the bright depressions may be places where lava has disintegrated into small fragments and collapsed, due to the mutual "popping" of trapped gas bubbles.

RMDSs are low mounds, almost all between 100 and 400 metres in diameter and between 3 and 8 metres high (Figure 6). Each one is surrounded by a depression – a moat – tens to hundreds of metres wide and a few metres deep. RMDSs are only found in lava-flooded areas, especially where the lava is rich in the element titanium, but not all such areas contain them. The simplest explanation for them is that they are places where lava from inside a flow has

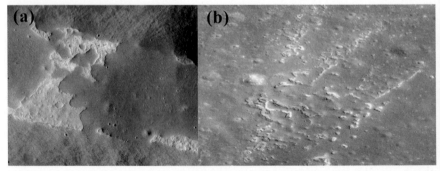

Figure 5. Irregular Mare Patches. (a) A large IMP on the floor of the graben Sosigenes. Part of NASA LRO image M177514916. Image is 2.25 kilometres wide. Dark lobate lava overlies bright fragmental material in a depression, possibly a vent, extending approximately horizontally across the middle of the image. (b) A cluster of small IMPs to the north of the impact crater Aristarchus, unrelated to any obvious volcanic vent. Image width is 825 metres. Part of NASA LRO image M168509312.

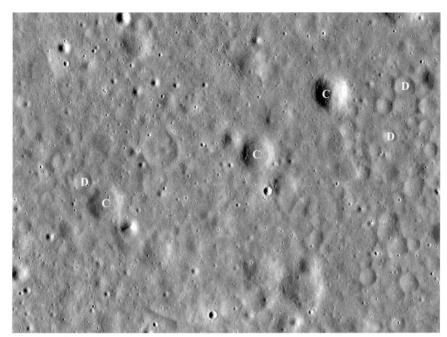

Figure 6. Ring-Moat Dome Structures. A large number of RMDSs, mostly about 250 metres in diameter, can be seen in the image. Letters C and D pick out a few examples of impact craters and domes to aid recognition. Part of NASA LROC image M1096293859. Image width is about 5 kilometres.

broken through the surface crust and oozed out. This can happen when a lava flow has stopped advancing due to excessive cooling, but the vent is still active, and forces fresh lava into the interior of the flow. The flow then inflates, and the stresses caused when different parts of the flow inflate by different amounts produce cracks in the surface from which the new lava can escape. Breakouts commonly occur on lava flows on Earth, but the resulting mounds rarely have shapes as regular as the lunar ones, a cause for much discussion.

The most controversial aspect of the discovery of IMPs and RMDSs concerns their ages. When the normal method of estimating the ages of lunar surface features – counting the numbers of small impact craters on their surfaces – is applied to these features, they appear to be mostly less than 100 million years old. All earlier measurements of the ages of lunar volcanic features, obtained from laboratory analyses of the Apollo rock samples and from counting the numbers of somewhat larger impact craters in pre-LRO images, gave ages greater than 1000 million years. This ten-fold difference in age would imply that the interior of the Moon was hot enough to drive volcanic eruptions very much more recently than previously thought, needing a major re-think of our understanding of the composition and structure of the lunar interior. There is, however, a possible way out of this conundrum. The absence of an atmosphere on the Moon, leading to the more efficient release of gas from lava, coupled with the low gravity which makes the pressure inside lava flows less than on Earth, mean that some lunar lavas should be foams, i.e., much more vesicular – full of gas bubbles – than those we see here. The way impact craters form in such materials is significantly different from how they form in solid rock, and if all of the craters produced in lava foam are smaller than in solid lava, which they should be, the age discrepancy can be explained.

The prediction that lunar lava flows were inflated implies that a single lava flow may consist of a series of layers: a dense layer at the top that has lost all of its gas in a lava-fountain, a very foamy layer under that, a complicated middle layer, a fourth layer that is foamy in some places but not in other places where the foam has escaped to the surface, and finally a dense layer at the bottom.

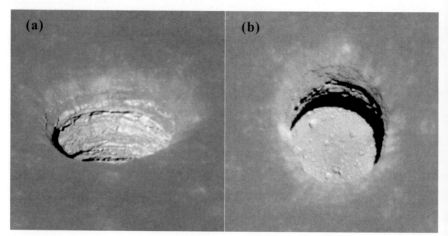

Figure 7. Views of (a) the walls (part of NASA LROC image M144395745) and (b) the floor (part of NASA LROC image M1315155623) of the ~125 metre deep, ~150 metre wide pit in the lava flows of Mare Tranquillitatis that is the target of the proposed Moon Diver mission. Layers visible in the walls may be multiple individual lava flow units or the internal structures of a smaller number of more complex flows.

Clearly what we need to check these predictions is a mission to the Moon that lets us probe down through the layers of lava flows in the mare basins to see what the flow interiors are really like. Just such a mission is currently being planned. Called Moon Diver, it would involve a soft-landing spacecraft sending a small robot to abseil into one of the deep pits seen in various places in the mare lavas, in this case into one in Mare Tranquillitatis (Figure 7). Equipped with cameras to map the walls of the pit, and instruments to measure the composition of the rocks, this would finally give us a complete 3-D picture of lunar lava flows.

Pages From the Past
Collecting Vintage Astronomy Books

Richard H. Sanderson

A cool evening breeze swept across the summit of Breezy Hill in Springfield, Vermont, as I sat with hundreds of stargazers and telescope makers. I was 20 years old and enjoying my fifth Stellafane convention back in 1975, but my life was about to take an unexpected turn.

As people glanced overhead, hoping to catch a Perseid meteor slicing through the darkening sky, keynote-speaker Kenneth Brown approached the microphone. Brown's enthusiasm was captivating as he told us about his hobby of collecting antique astronomy books. Hunting for relics of astronomy's past

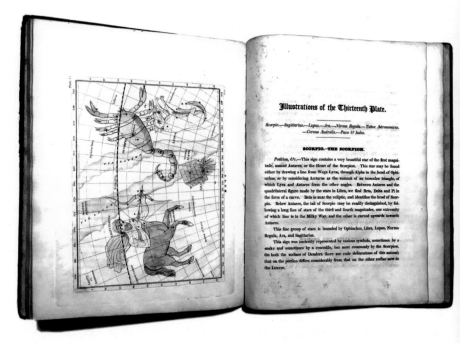

Early American star atlas: *Astronomical Recreations* by Jacob Green, Philadelphia, 1826.

seemed like lots of fun and during the months that followed, I took my first tentative steps into the arcane world of vintage astronomy books.

To understand why some people are driven to acquire old books filled with outdated information, we must explore the intrinsic value that these volumes possess. When I began plucking vintage astronomy books from the shelves of second-hand bookshops, I was fascinated by the archaic appearance of their covers and the quaint illustrations hiding inside. Gazing at their pages was like opening a window to a bygone epoch. Early astronomy books reveal footprints along a trail that has led us toward a more complete understanding the universe. Unlike new books that focus on history, vintage books place us in the present of an earlier time. As I read the words, I could almost hear the authors speaking to me from across the ages. Vintage books are tangible links to the past.

Some old books are endowed with artistic value. From hand-coloured mythological figures that adorn antique star atlases to striking engravings of giant telescopes and gorgeous chromolithographs of celestial scenery, many

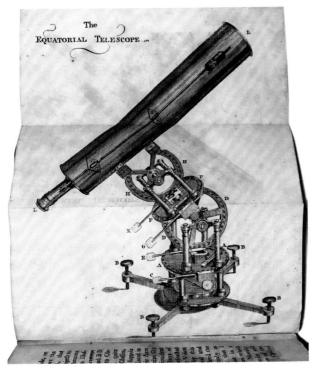

Foldout illustration from Volume 3 of *System of the Newtonian Philosophy, Astronomy, and Geography,* by Benjamin Martin, London, 1771.

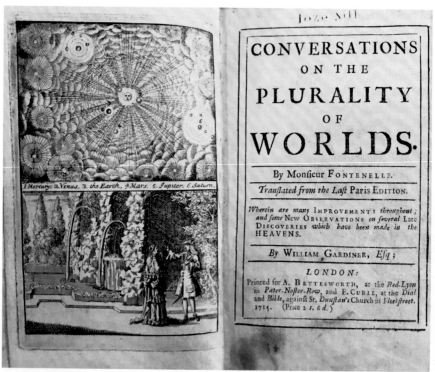

Conversations on the Plurality of Worlds by Bernard le Bovier de Fontenelle, translated by William Gardiner, London, 1715. Notice that the solar-system illustration only shows planets out to Saturn; Uranus was not discovered until 1781.

illustrations found in old astronomy books are visual treats. Finely-crafted leather bindings and beautiful marbled endpapers transform some early volumes into artistic masterpieces.

Old publications such as observatory reports and monographs can be valuable sources of information. One never knows when past observations will become important to modern research. With the shrinking of Jupiter's Red Spot, I recently found myself searching through my collection for early depictions of the giant planet. However, the informational value of old books has eroded in recent years as more and more scanned publications become available on the internet.

Some collectors are motivated by a desire to possess objects of great rarity, which explains why a wealthy collector would pay a fortune for a tiny piece of cardboard bearing the likeness of baseball player Honus Wagner. The acquisition of a rare astronomy book might require years of searching or perhaps a bit of incredible luck, but the hunt and ultimate capture of the prize will surely be an unforgettable adventure.

Some old books contain author inscriptions and autographs, or perhaps signatures or bookplates from important previous owners, any of which will enhance their historical value. Occasionally, a name inside an old book indicates that it once resided in the library of a renowned astronomer. I have found ownership signatures written by such notables as Percival Lowell, Henry Norris Russell, Elihu Thomson and Ambrose Swasey. What a unique privilege it is to hold in your hands a book that was owned and read by a famous astronomer.

If an author inscribes a copy of his or her book to another important person, that book becomes a very desirable double-association copy. It represents a link between two esteemed individuals. Carl Sagan was an important influence in my life and I'll always treasure my copy of the 1966 book *Intelligent Life in the Universe*, which Sagan inscribed "with admiration and best wishes" to Harvard astronomer Fred Whipple.

I'm always intrigued by the old names penned by unknown individuals that frequently appear inside vintage astronomy books. Each name offers the rewarding challenge of resurrecting the story of a forgotten life from the fog of time. An example is my 1853 edition of *Compendium of Astronomy*, written by Yale astronomer Denison Olmsted. It contains the pencil-written name of a young man from New Hampshire, Freeman Nourse, who survived the Civil War despite serving in a regiment that marched through swamps and was decimated by disease.

Another precious connection to the past emerged from three ownership names inside my copy of Benjamin Martin's *System of Newtonian Philosophy, Astronomy and Geography*, published in London in 1771. The book features gorgeous fold-out illustrations of telescopes and microscopes but those three names, two of them dated 1776, haunted my imagination. I discovered that they belonged to three Yale graduates who fought in the Revolutionary War. One of them, an officer named Ebenezer Huntington, participated in the Siege of Boston and was present at the surrender of Lord Cornwallis, which brought the war to a close. Huntington is depicted in John Trumbull's famous painting of that event, which hangs in the rotunda of the United States Capitol.

Where does a collector search for old astronomy books? Second-hand bookshops have long been a starting point for vintage-book lovers. Step inside and you will find anywhere from a handful of astronomy books in the science section to several shelves devoted to astronomy. The typical offering consists of newer second-hand books mixed with a spattering of common older volumes, punctuated by an occasional rare and collectible title. As your collection grows and your tastes become more refined, you will need to expand your search. Other sources of old astronomy books include antiquarian-book fairs and book auctions.

Illustration from *Tales About the Sun, Moon, and Stars* by Peter Parley (Samuel Griswold Goodrich), London, 1838.

One lesson that I've learned over the years is that a treasure can turn up in the unlikeliest of places. I'll never forget the day I found a rare 1926 first edition of *Amateur Telescope Making* in a second-hand bookshop and purchased it for a pittance. For me, that book symbolizes the birth of the telescope-making hobby in America. My copy of the Benjamin Martin book owned by the three Yale students came from a bookstore located in the basement of a restaurant!

Perhaps my oddest acquisition story dates from the early 1970s. I was walking past a trash dumpster when something caught my eye and I went over to investigate. Resting on a pile of office papers was a pristine, original copy of the *Apollo 11 Lunar Landing Mission Press Kit*. NASA distributed this 250-page publication two weeks before Neil Armstrong made humanity's first footstep on another world. What was temporarily a piece of trash has occupied a place of honour in my library for nearly half a century!

A couple years after I began collecting, an advertisement in *Sky & Telescope* alerted me to an antiquarian bookseller named Paul Luther who specialized in astronomy and was operating out of his home not far from where I lived. Unlike purveyors of second-hand books, antiquarian booksellers focus on older volumes and often specialize in specific types of books. The International League of Antiquarian Booksellers (**ILAB.org**) provides a directory on its website.

During the decades that followed, I eagerly perused Luther's catalogues and made periodic visits to browse the old and rare astronomy books that lined his shelves. Each visit was an adventure, but my greatest acquisition was the mentorship and personalized service that comes from dealing with an expert. Luther was largely responsible for transforming my budding hobby into a lifelong passion. He retired in 2013, due in part to the pressures exerted on book specialists by the internet.

The internet has transformed the landscape of the antiquarian-book world. Together with eBay, book-search sites like **Abebooks.com**, **Alibris.com** and **Biblio.com** enable collectors to hunt through tens of millions of old books from the comforts of home. Many professional booksellers list items for sale on these popular sites as well as on their own websites. I have found many wonderful books on **AddALL.com**, a useful website that combines more than 40 book-search sites and thousands of dealers into a single search engine.

Together with the convenience of hunting for old books on the internet comes an element of risk, but there are ways to mitigate that risk. When you

Fig. 28.—Lunar Day.

Illustration of the moon's surface from *The Atmosphere* by Camille Flammarion, English translation by C. B. Pitman, edited by James Glaisher, New York, 1873.

Total solar eclipse from the beautifully-illustrated book *Bilder-Atlas der Sternenwelt* by Edmund Weiss, Stuttgart, 1892.

purchase an item that you haven't examined, you must rely on the seller's description. Collectors should familiarize themselves with the antiquarian-book lexicon. An internet search will reveal useful glossaries like the one created by the Antiquarian Booksellers' Association of America (ABAA).

Not every internet seller is an expert and some are quite inexperienced. When considering the purchase of a vintage book through the internet, I recommend that you ask yourself the following questions. Does the seller have a high percentage of positive feedback from previous customers? Does the description tell you exactly what you are buying, including title, author, publication date, edition and type of binding? Is the book's condition carefully revealed? Is there a money-back guarantee if the book is not as described? When a bookseller examines dozens or hundreds of books for sale, a flaw like a missing illustration might slip through on rare occasions. Reputable dealers always offer a full refund when a defect has been inadvertently missed.

Packaging is another important concern. A book's corners are especially vulnerable to damage from rough handling during transit. Books should be packaged for shipment using a sturdy cardboard box and padding such as bubble wrap, so that a gap separates the corners of the book from the box. The majority of dealers make the effort to pack old books properly, but some inexperienced or lazy sellers simply insert a volume into a bubble-wrap mailing envelope, which often results in corner damage. Whenever I buy a book, I always request that it be packaged securely.

Collectors should strive to acquire the best copy of a book that is available and affordable. You may occasionally decide to relax your standards for exceedingly rare and expensive books. Purchasing an affordable copy in lesser condition might be preferable to never owning the book or waiting many years to find a perfect copy. On the other hand, I have never regretted those rare occasions when I exceeded my budget for a book that was in exceptionally fine condition or contained a coveted autograph. Most collectors avoid ex-library books due to the presence of unsightly markings and pockets, but the name of an observatory library is valuable provenance and adds to a book's interest.

What about first editions? A first edition, specifically the first printing of the first edition, often constitutes the earliest and most original version of an author's work. I say "often" because some books, such as *Astronomy With an Opera-Glass* (1888) by Garrett Serviss, are based on material that first appeared in serialized form in a journal or newspaper. A first edition may also be scarce compared to latter editions. Some books plainly state the edition number, while for others, a single date on the title page that matches a single date on the copyright page may be indicative of a first edition. However, the identification of first editions can sometimes become a challenging task. While I welcome first editions, especially for popular and important works, I do not insist that every book entering my collection fall into that category.

Collectors may gradually focus on one or more specific themes rather than dividing their efforts among the myriad topics that constitute the science of astronomy. Interesting subsets include telescope making, eclipses, women in astronomy, amateur astronomy, star and moon atlases, astronomical photography, children's astronomy books, local astronomy and specific authors. Focusing years of energy on a narrowly-defined theme can result in a unique and important collection, but my best advice is simply to collect whatever makes you happy.

The boundaries of a book collection can be expanded to include memorabilia and ephemera. Memorabilia are paper items other than books that were saved for their historical value, while ephemera are printed items meant to be used for a brief time and then discarded. Examples include old postcards, photographs, manuscripts, journals, telescope catalogues, lecture notices, broadsides, and the list goes on and on. I enjoy searching for early astronomy-related postcards as well as letters written by famous astronomers.

A book collection may ultimately become a treasured family heirloom, or perhaps it will one day be divided up and sold to other collectors or bequeathed to an institution. Over the decades, I have come to realize that I don't really own the volumes in my collection. Instead, I consider myself a custodian who enjoys the privilege of accompanying my books through time, but with that privilege comes responsibilities.

A good custodian will observe some simple precautions to assure that a cherished collection is protected from damage and deterioration, to assure that the books live on into the future. Books should be stored vertically and snugly in bookcases, but not so tight that extracting a volume becomes difficult. Old bindings may be somewhat fragile, so fight the temptation to hook your finger over the upper lip of a book's spine when removing it. Allowing books to lean over on a shelf or stacking heavy volumes atop one another can permanently distort old bindings.

Postcard, circa 1940, depicting the 40-inch refracting telescope at Yerkes Observatory.

Humidity poses a serious threat to books and other paper items. Very low humidity can cause brittleness, while high humidity promotes the growth of mould as well as foxing, which are those tan blemishes often seen on old paper. Ideally, vintage books should reside in a cool room where the humidity registers between 45% and 55%. Books should also be kept out of direct sunlight, which can bleach their spines. Attics and basements are unsuitable for book storage unless they are climate-controlled. Finally, common sense dictates that you keep books away from sources of airborne contaminants such as fireplaces, wood stoves and kitchens.

Old books, memorabilia and ephemera will benefit from acid- and lignin-free protective materials such as dust-jacket covers, postcard pages, boxes, tape and preservatives for leather bindings. These items can be purchased from archival-supply outlets.

The quest for old astronomy books will take each collector on a unique and rewarding intellectual journey, with many exciting detours along the way. A collector's knowledge of the history of astronomy will grow along with the collection, and being surrounded by old astronomy books will surely bring that history to life. If you decide to embark on your own journey through the universe of vintage astronomy books, as I did beneath the stars at Stellafane, I wish you the best of luck and happy hunting! I also invite you to visit my "Vintage Astronomy Books" Facebook group at **www.facebook.com/groups/vintageastronomy**

The Chances of Anything Coming from Mars
Martian Invasions That Never Happened

Jan Hardy

Giovanni Schiaparelli (1835–1910) was a respected Italian astronomer and science historian whose research varied extensively. A skilled observer, he studied the tail structures of comets, discovered the asteroid Hesperia and made many observations of the planet Mars even naming the "seas" and "continents" such as Tharsis, Chryse, Mare Australe and the Hellas basin, names which were adopted and are still in use today. During the Great Opposition of Mars in 1877 Schiaparelli observed what he reported as a network of linear structures on the Martian surface. He termed these structures "canali" meaning "channels", and suggested that these features were a means by which melt water from the polar ice caps could be distributed down to the dry equatorial region, water which could contain organic life.

Giovanni Schiaparelli, whose many contributions to astronomy included observations of the solar system, binary stars, the discovery of the asteroid 69 Hesperia, and proposed that meteor showers were associated with the tails of comets. (Wikimedia Commons)

An unfortunate mistranslation of "canali" into the English "canals" subsequently sparked off much speculation on their origin and construction and the possibility of life on Mars, as canals would imply artificial manufacture, possibly of an irrigation system, whereas channels would indicate natural geographical features. A number of respected astronomers, including the

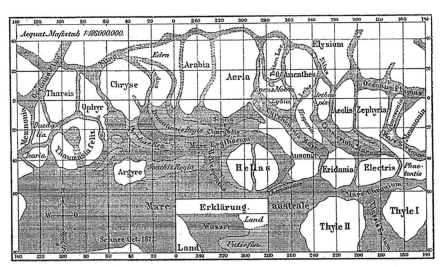

Giovanni Schiaparelli's map of Mars showing the "canali" (or "channels") he observed in 1877. His naming of the Martian "seas" and "continents" was based on classical geography (eg Chryse) or simply descriptive terms (eg Mare Australe, meaning Southern Sea). Most of these names are still in use today. (Wikimedia Commons / Meyers Konversations-Lexikon)

well-known American astronomer and leader in planetary astronomy Percival Lowell (1855–1916), supported this canal theory, Lowell himself spending much of his career attempting to discover signs of life on Mars using the 24-inch Clark telescope at the Lowell Observatory in Flagstaff, Arizona. Many other observers failed to see them. Numbered among these was Edward Emerson Barnard (1857–1923), who in 1916 discovered the highest known proper motion of a red dwarf star six light years from Earth in the constellation Ophiuchus (named Barnard's Star in his honour). Barnard attempted to observe the canals using the 36-inch telescope at Lick Observatory in California, but to no avail.

The idea of the existence of "Martians" captured the public imagination, and newspaper and magazine articles speculating on the nature of our supposed planetary neighbours abounded. Percival Lowell continued to be a strong advocate for the rest of his life and even published three books – *Mars* (1895), *Mars and its Canals* (1906), and *Mars as the Abode of Life* (1908), copies of which are still obtainable. However, in the years following Lowell's death the scientific community rejected the canal theory, and it was finally disproved in 1964 following the successful fly-by of Mars by NASA's Mariner 4 spacecraft, which

"A monstrous tripod, higher than many houses, striding over the young pine trees ..." An illustration by Henrique Alvim Corrêa, from the 1906 Belgium (French language) edition of *The War of the Worlds* by H. G. Wells. (Wikimedia Commons / Henrique Alvim Corrêa)

returned the very first images of another planet from space. These images revealed no evidence of canals anywhere on the surface. Its barren landscape and abundance of impact craters put paid to the idea that Mars could be inhabited, and the idea of structures created by intelligent Martians was finally laid to rest.

But by 1897, twenty years after Schiaparelli first reported seeing "canali", the idea of the existence of extra-terrestrials living on one of our planetary neighbours had settled into the public consciousness. It became a topic for all sorts of speculation and fiction, and not of all of it had these otherworldly inhabitants as benevolent and benign. One particular science fiction story was serialised by *Pearson's Magazine* in the UK and by *Cosmopolitan* magazine in the US, and told a gripping tale of how Martians invaded Earth in order to not only

Orson Welles (centre) at a meeting with reporters in November 1938 following his radio theatre broadcast of 'The War of the Worlds'. The broadcast had caused panic amongst many of its listeners. (Wikimedia Commons / The Express)

exploit the planet's resources but also to subjugate its human inhabitants and extract their blood as a source of nourishment. Of course, that story was *The War of the Worlds* by H. G. Wells, and was one of the earliest stories depicting a conflict between humans and extra-terrestrials. The narrative concludes with the defeat of the Martians by Earth-bound microbes to which the aliens had no immunity, luckily for us! The novel itself was first published in hardback a year later in 1898 and the story of tentacled Martian invaders was generally quite favourably received, despite its sometimes-violent depiction of the Martian's treatment of its human prey. The original setting of the novel was in and around Woking in Surrey, the home of H. G. Wells and his wife Catherine, but the invasion was relocated to Boston in the 1953 film 'The War of the Worlds' produced by George Pal and starring Gene Barry and Ann Robinson. Another version of the film, released in 2005 starring Tom Cruise and directed by Steven Spielberg, set the opening scenes in Bayonne, New Jersey.

And then of course there is the (infamous) occasion of 30 October 1938 when 'The War of the Worlds' was presented as a Halloween episode of the American radio drama anthology series 'The Mercury Theatre on the Air'. It was directed and narrated by the famous Orson Welles, whose baritone voice added gravity to the production. The story was presented as a normal evening of radio programming interrupted by a series of news bulletins reporting the "live" events as they happened, which purportedly caused panic amongst some of its audience who had perhaps tuned in part way through and had missed the show's introduction as a purely dramatic presentation. The actual extent of the panic has since been downplayed.

But 'War of the Worlds' did not end there! In 1978 the composer, musician and record producer Jeff Wayne released an album of rock opera using narration, progressive rock and a string orchestra to retell Wells' story. The album featured two discs, both roughly corresponding to two parts of the original novel – *The Coming of the Martians* and *The Earth under the Martians*. The album became a best-seller, selling 15 million copies worldwide.

The idea of invasions by extra-terrestrials became a recurrent theme in science fiction writing with the term "Martian" becoming vernacular for any being from outer space, not just those from the Red Planet. But Mars in particular has continued to fascinate and intrigue us and, perhaps, even scare us with its possibilities and proximity, and the fear of alien invasion has provided many an inspiration for novels, films and comics. The what-if scenario has endured!

An early example is *Last and First Men: A Story of the Near and Far Future*, a science fiction novel by British author Olaf Stapledon written in 1930. More of a past account than a direct prediction of our world two billion years in the future, it tells in retrospect the history of not just one occurrence of Martian aggression, but tens of thousands of years of war and invasion between Martians and humans – a disheartening depiction indeed!

In 1953 the movie 'Invaders from Mars' (remade in 1986) was the first feature film to depict aliens and their spacecraft in colour, and was rushed into production in order for it to beat George Pal's 'War of the Worlds' into theatres. In the film a young boy awakes one night to see a "flying saucer" disappear underground in a large pit behind his house. His father investigates but returns "changed", a fate soon to be followed by the boy's mother and neighbour. Mind-control devices have been implanted in their heads and the Martians use their victims in their attempt to sabotage a military project to build an atomic rocket. Scientists and the military together foil the dastardly plans and destroy the spaceship with explosives, and the film ends with the young boy back in his bed and once again being awoken by a loud noise and, on going to his window, sees the same flying saucer once more descending into the pit! However, the British release of the film had a different ending whereby the "dream versus reality" scenario is replaced with a more comforting final scene showing his parents, fortunately saved from their mind-control devices, tucking him into bed.

Feature films were not the only cinematic arena for tales of Martian invasions. During the 1950s cartoons were also a popular form of on-screen entertainment, and even Bugs Bunny faced a villainous man from Mars, making his debut in the 1948 Looney Tunes cartoon 'Haredevil Hare' along with his Martian dog K-9. The diminutive alien, dressed as an ancient Roman, became a popular character and yet remained without a name until 1979 when he became 'Marvin'. Although his appearances could not strictly be considered an 'invasion', the devious plots of 'Marvin the Martian' regularly revolved around attempts to destroy planet Earth.

Invasions and attacks from Mars were often a regular theme in comics, too. The story 'Martians, Go Home' by American science fiction writer Fredric Brown appeared in the September 1954 issue of *Astounding Science Fiction* and tells of a writer who witnesses an invasion of Earth by men from Mars, who in this case were stereotypically little and green. This description came into common use during the 1950s when reports of "flying saucers" started to become more widely publicised.

Trading cards, hugely popular in the 1950s and 1960s, were also not immune to our fascination with being invaded by creatures from outer space. In 1962, the New York based Topps Company produced a series of trading cards with a science fiction theme depicting an invasion by a race of merciless Martians intent on colonising Earth and making it their home world. The artwork on the cards, although colourful and in a comic book style, was often quite graphic in its depiction of the horror that ensued during the quite violent take-over of our planet. The Mars Attacks cards series became collectors' items, but at the time of their release during the early 1960s trading cards were popular with children. However, the sometimes quite graphic images caused a measure of public outrage, forcing Topps to temporarily halt their production. Twenty

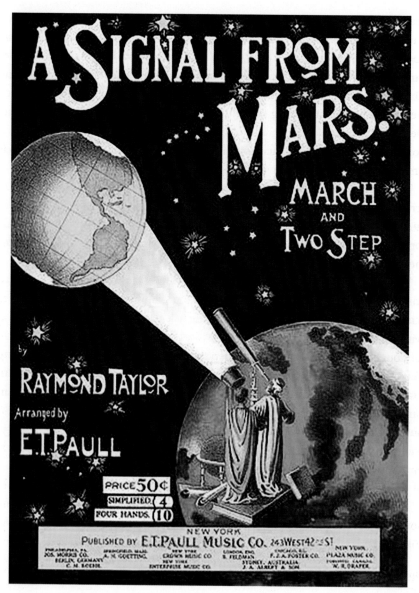

At around the same time that Percival Lowell was observing what he (wrongly) presumed to be canals on Mars (features first reported by Giovanni Schiaparelli in 1877) others were imagining their own vision of what intelligent life on Mars could look like. An example is seen on the cover for the sheet music *A Signal from Mars* published by Edward Taylor Paull, E.T.Paull Music Company, New York in 1901. (Wikimedia Commons/Raymond Taylor/Edward Taylor Paull/ E.T.Paull Music Company)

years later, in the 1980s, Topps issued a reprint of the original cards and added a range of Mars Attacks tie-in merchandise including the ever-popular comic books, but it was not until 1996 that a cinematic version based on the series came on the scene. Director Tim Burton brought the tale of Martian invasion to life in the comic science-fiction film 'Mars Attacks!' which featured a host of well-known actors including Jack Nicholson, Rod Steiger, Glenn Close, Danny DeVito and Annete Bening. Although the film received mixed reviews, with box-office takings somewhat less than what had been hoped for, the film has since achieved cult status.

The current plethora of superheroes in our cinemas and on our television screens are a testament to the enduring popularity of comic books since the 1930s, many characters not only battling super villains but also – and luckily for us – thwarting numerous attempted invasions by various alien species including Martians.

The examples above are just some of the many instances of our fictional conquest and sometimes destruction by more advanced alien species. Despite scientific evidence to the contrary, they illustrate not only an abiding fascination with, but also a perhaps irrational mistrust and fear of visitors from the unknown. Of course, not all fictional tales of alien invasion feature Martians, some having Venusians as our enemy and more still featuring unidentified life forms. An example is 'Invasion of the Body Snatchers', released in 1956 and based on an earlier science fiction novel by American science fiction writer Jack Finney. Yet our fascination with Mars and its possible inhabitants continues, and although we now know that, apart from ourselves, no other intelligent civilisation (hostile or otherwise) resides within our Solar System, we are still driven to find evidence of either current or past life on our planetary neighbours and their moons.

Explorations of the surface of Mars began in the early 1970s, first with the Soviet Mars landers (which had limited success) closely followed by NASA's Viking landers which provided the first panoramic views of the Martian

Opposite: NASA's Curiosity rover has found new evidence preserved in rocks on Mars that suggests the planet could have supported ancient life. A dusty self-portrait shows the rover at a drilled sample site called "Duluth" on the slower slopes of Mount Sharp, the central peak within Gale Crater, in 2018. Shallow test drills completed by Curiosity back in 2014 at a rock called "Windjana" were in preparation for collecting powdered sample material from the rock's interior. Could future sample return missions bring back more to Earth than expected? (Both images NASA/JPL-Caltech/MSSS)

surface in colour. Today the Mars Curiosity Rover (launched in 2011) continues to make discoveries as it investigates the planet's climate and geology. This includes looking for clues as to whether the environment has ever been suitable for harbouring microbial life, which may be buried deep below the Martian soil in places where liquid water once flowed in the distant past. Indeed, Curiosity has already identified the presence of ancient organic molecules (carbon, hydrogen) which although associated with life can also be created by non-biological processes, and are not necessarily indicators of the presence of life.

Although not a lander, the Mars Reconnaissance Orbiter (launched in 2005) transmits high-resolution images of the Martian surface back to Earth, and the recent discovery of signs of hydrothermal deposits on an ancient Martian sea-bed could provide hints as to how life on Earth originated. The ExoMars (Exobiology on Mars) mission, run jointly by the European Space Agency (ESA) and the Russian space agency Roscosmos, is designed to sample Martian material down to a depth of 2 metres and then analyse the samples using its on-board instruments. Manned exploration missions to Mars are still in the early planning stages and launch dates for the first humans to Mars are as yet a few years away, but in the meantime both NASA and ESA are proposing Mars sample return missions which would enable a more detailed scientific analysis of material from the Red Planet back here on Earth.

Of course, returning planetary material back to Earth has its own problems and hazards, not merely technical but also perhaps biological. After each Apollo moon landing mission, the astronauts went through a period of quarantine and decontamination, as did the samples of moon rocks they brought back with them, to ensure that they had returned to Earth with no unexpected passengers so to speak! Our scientific knowledge and mission experience have greatly increased since the days of Apollo, but despite our failure to find signs of life existing on any of our Solar System neighbours so far, any future manned or sample return missions to Mars would certainly require even more rigorous safety protocols to protect us from any form of contamination. As recounted above, it was our own Earthly microbes that ultimately defended us in H. G. Wells' fictional tale of Martian invasion. As we as a species step out further into the cosmos, who is to say alien microbes will not ultimately be our own undoing? But then … the chances of anything coming from Mars … ?

Māori Astronomy in Aotearoa-New Zealand

Pauline Harris, Hēmi Whaanga and Rangi Matamua

Māori are the Indigenous people of Aotearoa, New Zealand. Arriving from eastern Polynesian origins onboard large double hulled sailing vessels named *waka hourua*, the ancestors of the Māori utilised their knowledge of the sun, stars, moon and the environment to cross the large expanse of the Pacific ocean to settle Aotearoa-New Zealand around 1,280 AD (Lewis 1994; Prickett 2001). With this knowledge base they explored the breadth of their new home and established settlements throughout the country, adapting their island based system to new flora, fauna, geographical and climatic conditions. They re-storied and named their new home to remind them of their tropical Pacific origins (New Zealand Geographic Board 1990; Biggs 1991; Turoa 2000). Millennia of living with, studying and talking about the night sky were woven into the new sea, sky and landscape (Whaanga & Matamua 2016). Māori formulated a new language and embedded their extensive astronomical knowledge base across the breadth and depth of Māori society in oral forms such as *karakia* (incantations), *kōrero tuku iho* (oral tradition), *mōteatea* (traditional song), *whakataukī* (proverbs), in planting and harvesting practices and in the building of ancestral houses (Harris et al. 2013).

The sky to Māori represents the great sky father *Ranginui*, and the great earth mother *Papatūānuku*. From these two primal parents over 70 children were produced that presided over such realms as the forest, sea, winds and war (Smith 1913). The stars, moon, sun and planets also descend from these two beings (Matamua 2017a; Matamua 2017b; Best 1922), and Māori narratives describe the placement of the stars onto Rangi after the great separation of *Ranginui* and *Papatūānuku* that was undertaken by their children (ibid). After their separation, one of the children, *Tāne* adorned his parents; for their mother, he cloaked her with forests to keep her warm and their father, he adorned him with stars and other celestial beings. The celestial beings were obtained from his siblings *Tangotango* (the deep darkness of night) and *Wainui* (personification of the ocean) (ibid). The union of *Tangotango* and *Wainui* produced the 'Te Whānau Marama' the family of light (Whaanga & Matamua 2016). These

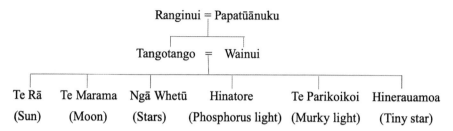

Figure 1: The genealogy of the celestial beings to their parents and grandparents *Ranginui* and *Papatūānuku*

genealogical relationships underpin all understanding and explanations of who the celestial beings are and how they relate to other beings on earth, including animals, plants, and humans, and the activities, practices and behaviours. Figure 1 shows the genealogy of the celestial beings.

Over the last 250 years, Māori knowledge, culture and language has been in significant decline. Since the arrival of the first British colonists, war, theft of land and resources, as well as introduced laws and legislation, such as the Native Schools Act and the Tohunga Suppression Act, have subjugated and inhibited the development and growth of Māori (Walker 2004). This, along with urban movement of Māori to larger cities over time, has seen a growing number of Māori becoming increasingly disconnected from their language, culture and knowledge. However, during the last 40 years great efforts have been undertaken to revitalise and regrow these, with Māori astronomy being one such system that is currently being revitalised.

This revitalisation of astronomical knowledge has seen much support from both Māori and non-Māori communities. This has manifested itself in the increased popularity of *Matariki* (Pleiades) celebrations (the Māori New Year) around the country which is signified by the heliacal rising of the Pleiades cluster. This acknowledgement has been largely due to the work of members of the Society for Māori Astronomy Research and Traditions (SMART) and in particular the works of Prof. Rangi Matamua whose seminal work called *Matariki Star of the Year* and his continued webshow has built large followings on Māori astronomy. Together with education programmes initiated by SMART and the voyaging community, these have seen Māori astronomy propelled into the minds and hearts of most New Zealanders.

In this paper we describe three key areas of Māori astronomy that continue to be developed in Aotearoa-New Zealand: the revitalisation of Māori astronomy and *Matariki*; the traditional calendar system called *maramataka*; and celestial navigation.

Rivalisation of Māori astronomy and Matariki

The interest in Māori astronomy has developed out of the remarkable growth in Māori-led celebrations and ceremonies of celestial events throughout Aotearoa-New Zealand (Hardy 2012; Matamua 2017a; Matamua 2017b). Leading the way in the revival of Māori astronomy is the growth and interest in celebrating the mid-winter heliacal rising of *Matariki*. The celebration of *Matariki* is becoming part of our national fabric with events, ranging from pre-dawn ceremonies, breakfasts, dinners, balls, and community days, being held throughout the country to foster community unity and togetherness. It has grown so much that it is now part of the national school curriculum, and its popularity is spreading amongst many city councils, government agencies and organisations, museums, social media, television, radio, print, and contemporary and traditional art forms (Whaanga & Matamua 2016). Connected with the growth in *Matariki* celebrations is the regeneration of a number of traditional ceremonies that coincide with the heliacal rising of *Matariki*: *Te taki mōteatea* (reciting of laments) and *Whāngai i te hautapu* (to feed with a sacred offering). The name *Matariki* is used to describe the entire star cluster with nine of the major stars in *Matariki* having their own individual names: *Tupuānuku* (Pleione), *Tupuānuku* (Atlas), *Waitī* (Maia), *Waitā* (Taygeta), *Waipunarangi* (Electra), *Ururangi* (Merope), *Pōhutukawa* (Sterope), *Hiwa-i-te-rangi* (Calaeno) and *Matariki* (Alcyone) (Matamua 2017a) (see Figure 2). Within the cluster each of these stars represents a specific purpose (i.e., either a food domain, or a weather occurrence) or a role intrinsically connected to the Māori world (i.e., the dead or the promise of a prosperous year). Although *Matariki* has been propelled as the star for the New Year, the star *Puanga* or Rigel is also used to signify the New Year for other tribes. *Puanga* celebrations thus occur around the country in the far north of the North Island, Taranaki and in parts of the South Island. *Matariki* ceremonies and celebrations occur after the rising of the *Matariki* cluster, in the last quarter moon phase (ibid). Figure 3 shows *Matariki* rising in the predawn sky with the moon above and the star *Puanga* also. In 2020 *Matariki* will be celebrated on 13 July and in 2021 on 2 July (ibid).

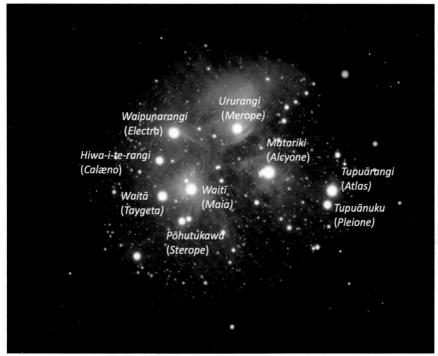

Figure 2: *Matariki* as observed in the southern hemisphere from Aotearoa-New Zealand and the Māori names associated with them. (*Matariki* image obtained and modified from Stellarium **stellarium.org**)

A number of research projects have played a key role in revitalising Māori astronomy. Two projects include *Te Mauria Whiritoi: Te Sky as a Cultural Resource* which was centred around cultural practices in Māori astronomy, and *Ngā Takahuringa ō te Ao: The Effect of Climate Change on Traditional Māori Calendars* that looked at the *maramataka* (Māori calendar) and how phenology (the study of cyclic and seasonal natural phenomena) embedded in the *maramataka* have been affected by climate change (see section below). Research from the *Te Mauria Whiritoi* project has fed into the development of the very successful web show *Living by the Stars* (**livingbythestars.co.nz**) and a publication of an English and Māori version of the book *Matariki, Star of the Year* by Prof. Matamua (2017a,b). The web show showcases different aspects of Māori astronomy and has over a million views to date. Other shows have also showcased Māori astronomy in the area of *maramataka* highlighting the works of Rereata Makiha, one of the leading experts on *maramataka* in Aotearoa-New Zealand.

Outreach has also played a key role in the revitalisation of astronomical knowledge. Over the past 30 years practitioners and academics have engaged with Māori communities, tribes and the general public in lectures, education programmes and museum exhibitions. This engagement has showcased and reinvigorated interest in Māori astronomy to the public including Māori and Pacific youth. Two major education programmes have been initiated by SMART: the *Tūhono i te Ao: Connecting the Worlds* programme that has become one of the most successful large scale Māori and Pacific Science programme run by Victoria University of Wellington's Te Roopu Awhina; the other was the *Te Reo Māori Planetarium* project that utilized a planetarium dome run by trainee navigators to deliver a Māori astronomy programme around the country to mainly Māori language schools (Harris 2017). These and other programmes have now reached over 10,000 Māori and Pacific youth and are regarded as some of the most successful outreach programmes in this space to date.

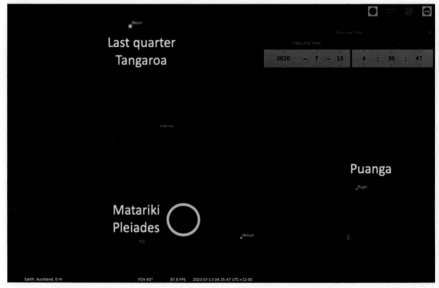

Figure 3: *Matariki* as observed when rising in the predawn sky with *Puanga* rising in the east and the moon in *Tangaroa* above. (Image obtained and modified from Stellarium **stellarium.org**)

Maramataka

The *Maramataka* is based on a system that incorporates time and the observations of phenomena during the year from the rising and setting of stars, to the phases of the moon, to the cycles of plants and animals, to what is occurring in our surrounding environment (Clarke and Harris 2017). One of the key components of the *maramataka* is the observations of the phases of the moon. For Māori there are 28 to 30 phases of the moon in the month, with each having a name that relates to a characteristic or relationship to that night / day (Roberts 2006; Tawhai 2013). With each phase there are associated observations and practices. Most commonly these are centred around but not limited to food hunting and gathering, where many *maramataka* refer to it being a good time to fish, hunt or plant certain species. *Maramataka* are regionally specific, with hundreds of *maramataka* existing following a similar framework but with locale specific differences and in some cases phase name variations (Roberts 2006; Tawhai 2013; Clarke and Harris, 2017). An example of some nights of a *maramataka* is shown in the table below from the tribe Ngāti Kahungunu (Mitchell 1972), which is located on the east coast of the North Island of Aotearoa-New Zealand. Here we see the different phase names along with their descriptions. Although seemingly simplistic in table form, a great deal of knowledge and layers of information goes behind these descriptors, which is not commonly written down but were retained in memory by practitioners.

Index of Night	Name of Night	Relevant Description of the Night
1	Whiro	Bad day; the moon is out of sight
2	Tirea	Bad day; the moon is slightly seen (New Moon)
3	Hoata	Good day for planting and fishing, the moon is well shown
13	Mawharu	Good day, especially for pot cray-fishing
16	Rakanui	Good day for planting, but not for fishing (Full Moon)
24	Tangaroa-amua	Good day for deep sea fishing; white bait is running

Some of the moon nights for the *maramataka* for the tribe Ngāti Kahunungu (Mitchell 1972).

Another key component of the *maramataka* is the rising and setting of particular stars that indicated certain times of the year. As described earlier stars such as *Matariki* have been propelled in popularity in communities around Aotearoa-New Zealand. *Matariki* and other stars such as *Puanga* (Rerekura 2014), *Whānui* (Best 1922) and *Pipiri* (Matamua, 2017a) were used to indicate certain times of the year. The heliacal rising of stars was typically used for determining the time of year for rituals, harvesting, planting etc. For example stars such as *Whānui* (Vega) indicated the time for harvesting of the *kūmara* (sweet potato – *Ipomoea batatas*), which typically occurs in the Gregorian calendar time of February/ March. The rising of *Whānui* occurs around early March in Aotearoa-New Zealand where it will be observable in the early pre-dawn sky. Traditional Māori narrative describes how the *kūmara* was gained from its celestial origins and brought to earth where it currently dwells. The story involves *Whānui* (the father of the kūmara), his brother *Rongomaui* and his wife *Pani-tinaku*. *Rongomaui* ascended to heavens to obtain the offspring of *Whānui* and unable to persuade his brother to give up his children, he stole *Whānui's* children – this being the origin of theft. On his return back to earth he dwelled with his wife *Pani-tinaku* who gave birth to nine children – the different varieties of *kūmara*. In punishment for his theft, *Whānui* annually sends down his other children the caterpillars and moths to destroy the offspring of his brother (Best 1922). There are many narratives such as this that relates the times in the *maramataka* to the timing of agricultural, fishing, hunting activities, and rituals.

Over the past six years the popularity and interest of *maramataka* has increased dramatically. Researchers and Māori communities are now engaging in research to grow knowledge of their calendar system. A range of projects are implementing astronomical knowledge and how it can inform phenological observations of plants and animal cycles, climate change, mental health and well-being. These modern interpretations are part of the growing movement to regain knowledge in a modern context to monitor the health and well being of us as humans but also the world around us (Whaanga, Matamua and Harris, in press).

Voyaging and Celestial Navigation

Māori have voyaged for millennia across the vast expanses of the ocean to populate the largest region of the planet utilizing observations of the sun, moon, stars, swells, winds and other indicators (Matamua, Harris & Kerr 2013). This voyaging and its methods have captured the imaginations of people worldwide. Over 40 years ago the likes of Hector Busby and Nainoa Thompson began to revitalize these traditions under the tutelage of Master navigator Mau Piailug from Satawal in the Caroline Islands. From this began the renaissance of oceanic voyaging for Hawai'i, Aotearoa-New Zealand and other islands in the

Pacific. As part of this renaissance, celestial navigation for Māori and Hawaiian navigators has been adapted from knowledge passed down by Mau Piailug. The techniques and observations are based on a compass system that has 32 sections called houses (Te Aurere Education Package 1995). These houses are located on the horizon and each house is named. Out of each house the planets, stars, sun and moon rise and set on the opposite side and is used to determine direction when sailing.

As one sails to the north or south, new stars will be appear above the horizon. For example if you travel north new stars will be seen towards the north and the stars to the south will drop below the horizon. Thus, certain stars can be used to tell certain latitudes where known islands are. Therefore, the sky will act as a map to guide navigators to their island destination. Latitude is much easier to estimate than longitude which needs other knowledge and experience to utilise in order to reach your destination. Zenith stars (the stars that are directly above you) can also be used to indicate latitude (Matamua et al. 2013). However, in practicality, on a moving *waka* or sailing vessel it can be challenging to see what is directly up, thus the position of stars relative to the horizon might be better to use to tell latitude. In order to determine north or south, navigators today learn at least 60 pairs of stars. These stars are called meridian pairs and are stars that cross the meridian at the same time and can be used to determine a north-south line and from this the determination of all other cardinal points or houses.

Thus far we have discussed how stars can be used to tell direction which only applies when it is dark. During the day the sun is fundamental in setting direction when voyaging. This is primarily done at sunrise and sunset whereas at night many stars can be used to set your direction. Although stars are used as a definite indicator for direction its usefulness relies heavily on the weather, in particular cloud cover, which can hinder observations significantly at sea. As a result, a trained navigator relies on more than just the sun, stars and moon; they also have to be adept in the observation of swells and currents of the ocean, the winds, and animals together with inherent knowledge to guide them in where they need to go. Today voyaging has become an integral part of Māori identity and cultural revitalisation. The work done by the likes of Busby, Hoturoa Kerr and Jack Thatcher and many other experts in the field has had a significant impact on Māori astronomy, culture and society in addition to playing a critical role in reconnecting the Pacific nations together under the voyaging movement.

Conclusion

Māori astronomy over the last 30 years has grown into the consciousness of the nation in Aotearoa-New Zealand. This body of knowledge and all it encompasses are steeped in its origins from the primal parents of *Ranginui* and *Papatūānuku* through to everything in the known world and universe. This genealogical relationship manifests in how we relate the sky to the earth, and the earth to the sky, through observations and understandings in the *maramataka*, to the language, rituals and practices that honour the celestial canopy. Māori astronomy includes a great depth and inter-related understandings to more than the sky, but connects to what is occurring in the environment, in our ecological processes and indeed our own health and well-being. The efforts to revitalise Māori astronomy has seen practitioner development (Whaanga et al. in press), museum exhibitions and education outreach across Aotearoa-New Zealand. National celebrations acknowledging Māori astronomy are now bringing New Zealanders together as our national identity becomes more inclusive of Māori knowledge to grow more into a bicultural nation. As knowledge is built upon we also see the application of Māori astronomical knowledge into new areas such as space exploration and the application to areas such as climate change and sustainability. The future of Māori astronomy bodes well with a growth in practitioners, high quality research, and the delivery of a growing online presence through web-series. It is safe to say that as Māori astronomy continues to grow strong, it will set a new course towards a horizon guided by the stars of our ancestors.

Bibliography

Best, E., 1922, *The Astronomical Knowledge of the Maori, Genuine and Empirical: Including Data Concerning Their Systems of Astrogeny, Astrolatry, and Natural Astrology, with Notes on Certain Other Natural Phenomena*, Govt. Print., Wellington, New Zealand.

Biggs, B. 1991, 'A Linguist Revisits the New Zealand Bush', In: Pawley A, Editor. *Man and a Half: Essays in Pacific Anthropology and Ethnobiology in Honour of Ralph Bulmer*. Auckland: Polynesian Society; p. 67–72.

Clarke, L. and Harris, P., 2017, 'Maramataka', in H. Whaanga, T. T. Keegan & M. Apperley (ed.), *He Whare Hangarau Māori – Language, culture & technology* [Ebook], Te Pua Wānanga ki te Ao, Te Whare Wānanga o Waikato, Hamilton, New Zealand, pp. 129–135.

Hardy, A., 2012, 'Re-designing the national imaginary: the development of Matariki as a contemporary festival', *Australian Journal of Communication*, vol. 39, no. 1, pp. 103–119.

Harris, P., 2017, 'Portable planetariums in the teaching of Māori astronomy', in H. Whaanga, T. T. Keegan & M. Apperley (ed.), *He Whare Hangarau Māori – Language, culture & technology* [Ebook] Te Pua Wānanga ki te Ao, Te Whare Wānanga o Waikato, Hamilton, New Zealand, pp. 136–148.

Harris, P., et al. 2013, 'A review of Māori Astronomy in Aotearoa-New Zealand', *The Journal of Astronomical History and Heritage*, vol. 16, no. 3, pp. 325–336.

Lewis, D., 1994, *We, the Navigators: the Ancient Art of Landfinding in the Pacific*, Honolulu: University of Hawai'i Press.

Prickett, N., 2001, *Māori Origins from Asia to Aotearoa*, Auckland: Auckland Museum.

Turoa, T., 2000, *Te Takoto o Te Whenua o Hauraki: Hauraki Landmarks*, Auckland: Reed.

Matamua, R. L., et al. 2013, 'Māori Navigation', in (ed.), *New Zealand Astronomical Society Yearbook 2013*, Stardome Observatory Planetarium, Auckland: pp. 28–34.

Matamua, R., 2017a, *Matariki: Te Whetū Tapu o Te Tau, Huia*, Wellington, New Zealand.

Matamua, R., 2017b, *Matariki – The Star of the Year, Huia*, Wellington, New Zealand.

Mitchell, J. H., 1972, *Takitimu*, Southern Reprints, Auckland, New Zealand.

New Zealand Geographic Board. 1990, *He Kōrero Pūrākau Mo Ngā Taunahanahatanga a Ngā Tūpuna: Place Names of the Ancestors, a Māori Oral History Atlas*, Wellington.

Rerekura, S., 2014, *Puanga: Star of the Māori New Year*, Te Whare Wananga o Ngapuhi-nui-tonu, Auckland, New Zealand.

Roberts, M., et al. 2006, *Maramataka: the Māori Moon Calendar*, AERU, Canterbury, N.Z.

Tāwhai, W., 2013, *Living by the Moon: Te Maramataka a Te Whānau-ā-Apanui*, Huia, Wellington, New Zealand.

Smith, S. P., et al. 1913, *The Lore of the Whare-waananga, or, Teachings of the Maori College on Religion, Cosmogony and History*, Vol 1: Te Kauwae-runga, or 'things celestial', printed for the Society by T. Avery, New Plymouth, New Zealand.

Te Aurere Education Package, Rarotonga –Aotearoa 1995 Voyage.

Walker, R., 2004, *Ka Whawhai Tonu Matou: Struggle without End* (1990).

Whaanga, H. & Matamua, R. 2016, 'Matariki Tāpuapua: Pools of Traditional Knowledge and Currents of Change', in M. Robertson and P. K. E. Tsang (ed.), *Everyday Knowledge, Education and Sustainable Futures: Transdisciplinary Research in the Asia/Pacific Region*, Singapore, Springer, pp. 59–70.

Whaanga, H., Matamua, R. and Harris, P., (in press), 'The Science and Practice of Māori Astronomy and Matariki', Wellington, *New Zealand Science Review*.

Miscellaneous

Some Interesting Variable Stars

Tracie Heywood

You may have considered taking up variable star observing but how should you choose which stars to observe? There are so many variable stars in the night sky and you don't want to waste your time attempting to follow the "wrong" ones. Your choice of stars will, of course, depend on the equipment that you have available, but also needs to be influenced by how much time you can set aside for observing.

This article splits some of the more interesting variable stars into three groups. The group that is most suited to you will be determined by how often you can observe each month and for how long you can observe on a clear night. The light curves included have been constructed from observations stored in the BAA Photometry Database. Comparison charts for all of these stars can be found on the British Astronomical Association Variable Star Section website at **www.britastro.org/vss**

One-Nighters

These are stars that can go through most of their brightness variations in the course of a single (reasonably long) night and would suit *people who can only observe occasionally, but can then observe well into the night.*

STAR	TYPE	RA		DEC		MAX / MIN	PERIOD
		H	M	°	′		
RZ Cassiopeiae	EA	02	49	+69	38	6.2 / 7.7	1.195255 days (~29 hours)
U Geminorum	UGSS+E	07	55	+22	00	8.2 / 14.9	4.25 hours for eclipses
IP Pegasi	UG+E	23	23	+18	25	12.0 / 18.6	3.80 hours for eclipses
W Ursae Majoris	EW	09	44	+55	57	7.8 / 8.5	0.336637 days (~8 hours)
Beta (β) Persei	EA	03	08	+40	57	2.1 / 3.4	2.86736 days

RZ Cassiopeiae is a particularly good star for beginners who use binoculars. It has a good-sized brightness range and its eclipses, taking less than five hours, are shorter than for most other eclipsers. Predictions for upcoming eclipses can be found at **www.as.up.krakow.pl/minicalc/CASRZ.HTM**

U Geminorum is usually followed by observers as a dwarf nova that produces several outbursts each year from its 14th magnitude base level to around 9th magnitude. At maximum, it is thus within the realm of larger binoculars or small telescopes. Of relevance here, however, is that U Gem is also a very close binary system which produces eclipses. These can be seen superimposed on the light curve of the outbursts.

IP Pegasi is another dwarf nova system that shows eclipses. However, since it only reaches around 12th magnitude during outbursts, a somewhat larger telescope is required than for U Geminorum.

Predictions for eclipses of U Geminorum and IP Pegasi and other similar stars can be found at **www.britastro.org/vss/CV_eclipse_predictions.htm**

W Ursae Majoris is an eclipsing binary system, containing two almost identical stars that are in contact with each other. This means that as soon as one eclipse ends, the next one starts. Each eight hour cycle includes two eclipses. Predictions for upcoming eclipses can be found at **www.as.up.krakow.pl/minicalc/UMAW.HTM**

Beta (β) Persei (Algol) is an eclipsing variable whose changes can be detected with the naked eye, although its 9.6-hour long eclipses are rather long to observe in full.

Most Clear Nights

These are stars that vary a bit more slowly, but which can display significant changes over a week or two. They would suit *people who can observe for a short while on (nearly) every clear night.*

STAR	TYPE	RA		DEC		MAX / MIN	PERIOD
		H	M	°	′		
CH Cygni	ZAnd+SR	19	25	+50	14	5.6 / 10.1	97 days, 725 days
SS Cygni	UGSS	21	43	+43	35	7.7 / 12.4	None
Beta (β) Lyrae	EB	18	50	+33	18	3.3 / 4.3	12.94187 days
Z Ursae Minoris	RCB	15	02	+83	03	10.8 / 19.0	None

CH Cygni is a 'symbiotic' variable. It is a binary system consisting of a hot star showing erratic variations and a late type star showing periodic variations. The brightness changes are often largely periodic, but at other times, these are "swamped" by the more erratic variations of the hotter star. CH Cygni was particularly bright during the early 1980s and rather faint during the mid-1990s.

SS Cygni is a dwarf nova that spends most of its time at 12th magnitude, but every 7-8 weeks brightens dramatically by 3-4 magnitudes before fading again over the next fortnight.

Beta Lyrae is an eclipsing variable in which the two stars have gravitationally distorted each other. As a result, brightness variations continue outside of the eclipses. Mass transfer between the two stars causes the orbital period of the two stars to gradually increase. As a result, observed eclipse times gradually move out of step with predictions based on out-of-date values for the period.

Z Ursae Minoris, located quite close to the north celestial pole, is a similar type of star to R Coronae Borealis in that it shows deep unpredictable fades. At maximum, it is of 11th magnitude, but typically drops by around 5 magnitudes during the fades.

Several Times per Month

Slower variables whose brightness will change significantly over several months or a year. These variables would suit *people who can observe several times per month, but not necessarily on every clear night.*

STAR	TYPE	RA		DEC		MAX / MIN	PERIOD
		H	M	°	′		
T Cephei	Mira	21	10	+68	29	5.4 / 11.0	389 days
AF Cygni	SR	19	30	+46	09	6.2 / 7.9	93 days
TX Draconis	SR	16	35	+60	28	6.8 / 8.2	77 days
U Monocerotis	RVB	07	31	−09	47	5.5 / 7.7	91.3 days
R Scuti	RVA	18	47	−05	42	4.2 / 8.6	146.5 days
T Ursae Majoris	Mira	12	36	+59	29	6.6 / 13.5	257 days
RZ Vulpeculae	RVB	19	47	+19	29	11.7 / 15.0	89.5 days

T Cephei is one of the brightest Mira-type variables. As for other stars of this type, the brightness at maximum varies between cycles and can sometimes be up to a magnitude fainter than listed here. Most of T Cephei's brightness range can be followed using binoculars, although a small telescope is required to follow it when close to minimum. Having passed through minimum in December 2020, it will then brighten towards a late summer maximum.

AF Cygni and **TX Draconis** are semi-regular variables. "Semi-regular" indicates that the brightness variations only roughly repeat from one cycle to the next. The listed period is only an average and the observed brightness range will differ between cycles, sometimes covering the whole of the listed range, but often being somewhat smaller.

U Monocerotis and **R Scuti** are RV Tauri type variables. These are pulsating stars that often show alternate shallow and deep minima over a period of several months. R Scuti belongs to the RVA sub-category in which the average long-term brightness of the star remains constant. U Monocerotis in contrast belongs to the RVB sub-class in which the average brightness shows longer term changes lasting several years.

T Ursae Majoris is a Mira-type variable which has an average period of approximately 8.5 months. In early 2021 it will be fading from a December maximum, but will brighten again during the summer as it approaches an early September maximum.

RZ Vulpeculae is an RV Tauri type variable of the RVB subtype. The 8-year timescale of the accompanying light curve is obviously too long for minima related to the 89.5-day period to be seen, but it does bring out the longer "supercycle" variations with a period of around four years.

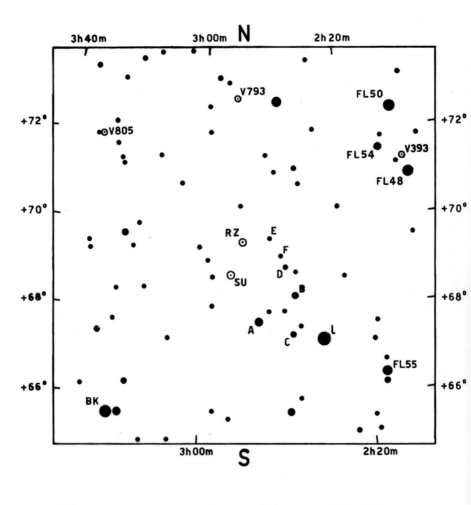

The BAA VSS finder chart for RZ Cassiopeiae, with six suitable comparison stars labelled A–F. (BAA Variable Star Section)

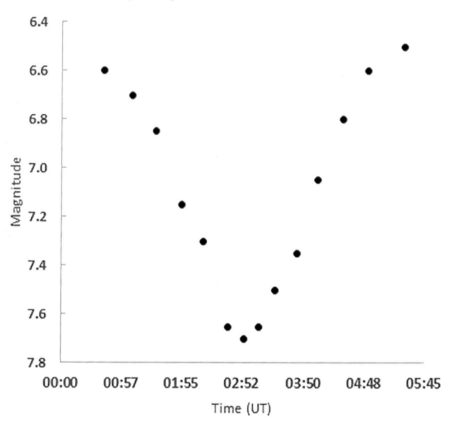

Light curve for the primary eclipse of RZ Cassiopeiae based on observations made by Tracie Heywood using 11x80 binoculars) during the early hours of January 8th 2018. (Tracie Heywood)

Light Curve for CH CYG

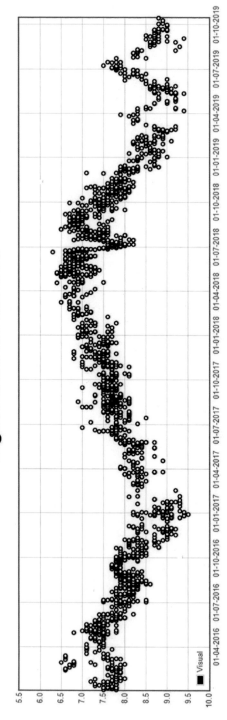

Light curve for CH Cygni between January 2016 and summer 2019. (BAA Photometry Database)

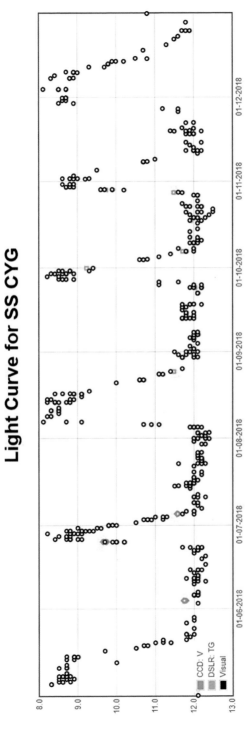

Light curve for the dwarf nova SS Cygni between May and December 2018. (BAA Photometry Database)

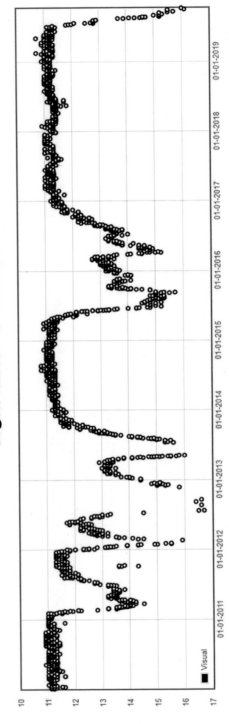

Light curve for Z Ursae Minoris between January 2010 and autumn 2019. (BAA Photometry Database)

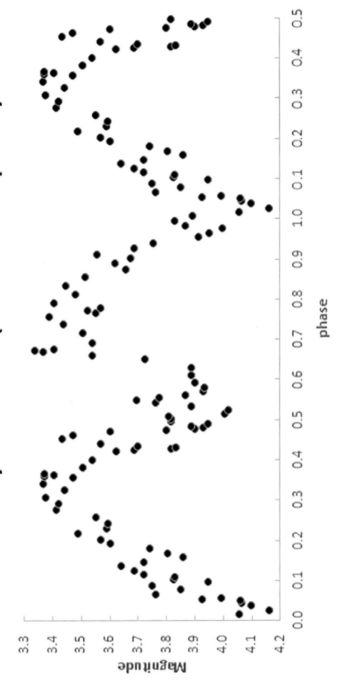

Light curve for the eclipsing variable Beta (β) Lyrae in which all observations reported during 2016 have been combined into a single light curve (showing 1.5 cycles of its 12.94-day period). (BAA Photometry Database / Society for Popular Astronomy Variable Star Section)

Light Curve for T CEP

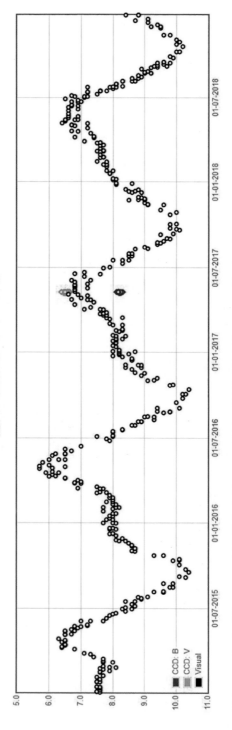

Light curve for the Mira-type variable T Cephei between 2015 and 2018. (BAA Photometry Database)

Light Curve for AF CYG

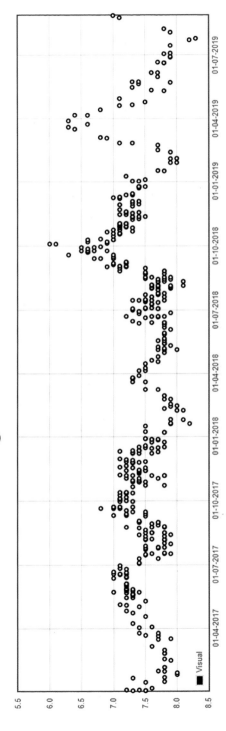

Light curve for the semi-regular variable AF Cygni between January 2017 and summer 2019. (BAA Photometry Database)

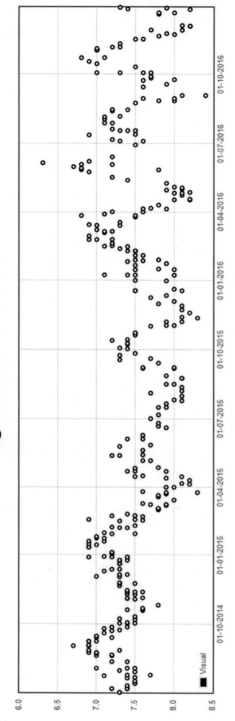

Light curve for the semi-regular variable TX Draconis between July 2014 and December 2016. (BAA Photometry Database)

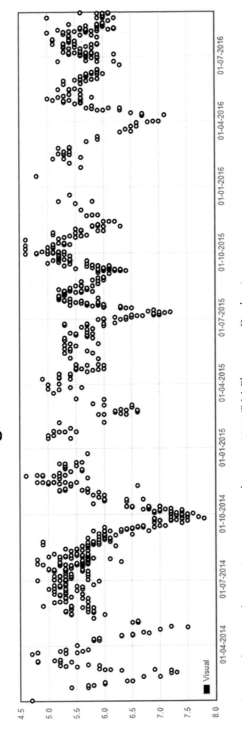

Light curve for R Scuti between January 2014 and summer 2016. (BAA Photometry Database)

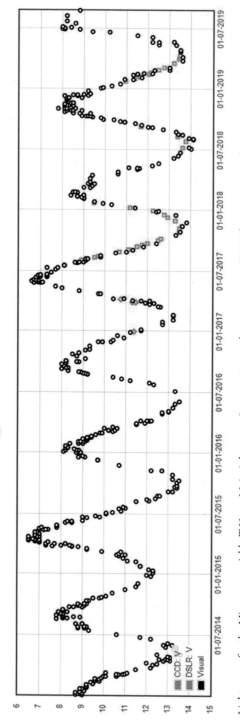

Light curve for the Mira-type variable T Ursae Majoris between January 2014 and summer 2019. (BAA Photometry Database)

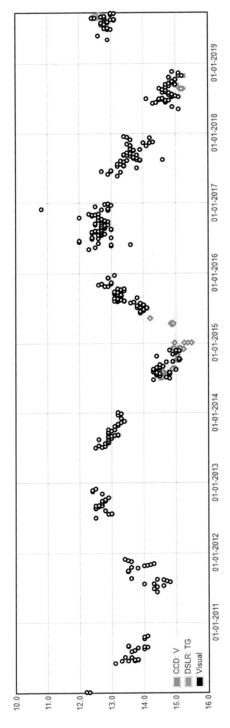

Light curve for RZ Vulpeculae between January 2014 and summer 2019. (BAA Photometry Database)

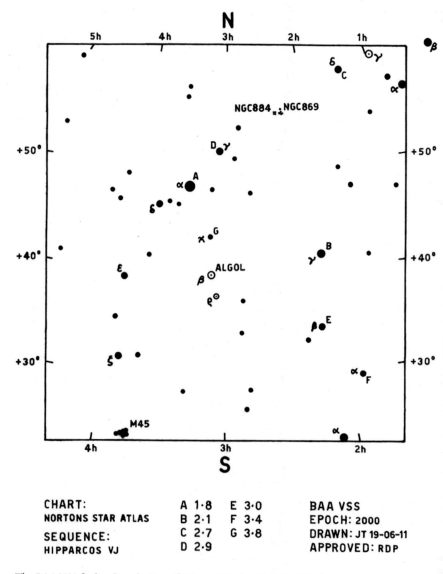

The BAA VSS finder chart for Beta (β) Persei (Algol). (BAA Variable Star Section)

Minima of Algol in 2021

Beta (β) Persei (Algol): Magnitude 2.1 to 3.4 / Duration 9.6 hours

		h				h				h				h	
Jan	3	19.9		Feb	1	12.1		Mar	2	4.2		Apr	2	17.2	
	6	16.7			4	8.9			5	1.0	⋆		5	14.0	
	9	13.5			7	5.7			7	21.8	⋆		8	10.8	
	12	10.3			10	2.5	⋆		10	18.7			11	7.6	
	15	7.2			12	23.3	⋆		13	15.5			14	4.4	
	18	4.0	⋆		15	20.1	⋆		16	12.3			17	1.3	
	21	0.8	⋆		18	16.9			19	9.1			19	22.1	
	23	21.6	⋆		21	13.8			22	5.9			22	18.9	
	26	18.4			24	10.6			25	2.7	⋆		25	15.7	
	29	15.2			27	7.4			27	23.5	⋆		28	12.5	
									30	20.4					

		h				h				h				h	
May	1	9.3		Jun	1	22.3		Jul	3	11.3		Aug	1	3.4	
	4	6.2			4	19.1			6	8.1			4	0.2	
	7	3.0			7	15.9			9	4.9			6	21.1	
	9	23.8			10	12.8			12	1.7			9	17.9	
	12	20.6			13	9.6			14	22.5			12	14.7	
	15	17.4			16	6.4			17	19.4			15	11.5	
	18	14.2			19	3.2			20	16.2			18	8.3	
	21	11.0			22	0.0			23	13.0			21	5.1	
	24	7.9			24	20.8			26	9.8			24	2.0	⋆
	27	4.7			27	17.6			29	6.6			26	22.8	
	30	1.5			30	14.5							29	19.6	

		h				h				h				h	
Sep	1	16.4		Oct	3	5.4		Nov	3	18.3		Dec	2	10.5	
	4	13.2			6	2.2	⋆		6	15.2			5	7.3	
	7	10.0			8	23.0	⋆		9	12.0			8	4.1	
	10	6.9			11	19.8			12	8.8			11	1.0	⋆
	13	3.7			14	16.6			15	5.6			13	21.8	⋆
	16	0.5	⋆		17	13.5			18	2.4	⋆		16	18.6	
	18	21.3			20	10.3			20	23.2	⋆		19	15.4	
	21	18.1			23	7.1			23	20.1			22	12.2	
	24	14.9			26	3.9	⋆		26	16.9			25	9.0	
	27	11.7			29	0.7	⋆		29	13.7			28	5.8	
	30	8.6			31	21.5							31	2.7	⋆

Minima marked with an asterisk (⋆) are favourable from the British Isles, taking into account the altitude of the variable and the distance of the Sun below the horizon (based on longitude 0° and latitude 52°N)

times given in the above table are expressed in UT/GMT.

Some Interesting Double Stars

Brian Jones

The accompanying table describes the visual appearances of a selection of double stars. These may be optical doubles (which consist of two stars which happen to lie more or less in the same line of sight as seen from Earth and which therefore only appear to lie close to each other) or binary systems (which are made up of two stars which are gravitationally linked and which orbit their common centre of mass).

Other than the location on the celestial sphere and the magnitudes of the individual components, the list gives two other values for each of the double stars listed – the angular separation and position angle (PA). Further details of what these terms mean can be found in the article *Double and Multiple Stars* published in the 2018 edition of the Yearbook of Astronomy.

Double-star observing can be a very rewarding process, and even a small telescope will show most, if not all, the best doubles in the sky. You can enjoy looking at double stars simply for their beauty, such as Albireo (β Cygni) or Almach (γ Andromedae), although there is a challenge to be had in splitting very difficult (close) double stars, such as the demanding Sirius (α Canis Majoris) or the individual pairs forming the Epsilon (ε) Lyrae 'Double-Double' star system.

The accompanying list is a compilation of some of the prettiest double (and multiple) stars scattered across both the Northern and Southern heavens. Once you have managed to track these down, many others are out there awaiting your attention …

Star	RA h	RA m	Declination °	Declination ′	Magnitudes	Separation (arcsec)	PA °	Comments
Beta[1,2] (β[1,2]) Tucanae	00	31.5	−62	58	4.36 / 4.53	27.1	169	Both stars again double, but difficult
Achird (η Cassiopeiae)	00	49.1	+57	49	3.44 / 7.51	13.4	324	Easy double
Mesarthim (γ Arietis)	01	53.5	+19	18	4.58 / 4.64	7.6	1	Easy pair of white stars
Almach (γ Andromedae)	02	03.9	+42	20	2.26 / 4.84	9.6	63	Yellow and blue-green components
32 Eridani	03	54.3	−02	57	4.8 / 6.1	6.9	348	Yellowish and bluish
Alnitak (ζ Orionis)	05	40.7	−01	57	2.0 / 4.3	2.3	167	Difficult, can be resolved in 10cm telescopes
Gamma (γ) Leporis	05	44.5	−22	27	3.59 / 6.28	95.0	350	White and yellow-orange components, easy pair
Sirius (α Canis Majoris)	06	45.1	−16	43	−1.4 / 8.5			Binary, period 50 years, difficult
Castor (α Geminorum)	07	34.5	+31	53	1.93 / 2.97	7.0	55	Binary, 445 years, widening
Gamma (γ) Velorum	08	09.5	−47	20	1.83 / 4.27	41.2	220	Pretty pair in nice field of stars
Upsilon (υ) Carinae	09	47.1	−65	04	3.08 / 6.10	5.03	129	Nice object in small telescopes
Algieba (γ Leonis)	10	20.0	+19	50	2.28 / 3.51	4.6	126	Binary, 510 years, orange-red and yellow
Acrux (α Crucis)	12	26.4	−63	06	1.40 / 1.90	4.0	114	Glorious pair, third star visible in low power
Porrima (γ Virginis)	12	41.5	−01	27	3.56 / 3.65			Binary, 170 years, widening, visible in small telescopes
Cor Caroli (α Canum Venaticorum)	12	56.0	+38	19	2.90 / 5.60	19.6	229	Easy, yellow and bluish
Mizar (ζ Ursae Majoris)	13	24.0	+54	56	2.3 / 4.0	14.4	152	Easy, wide naked-eye pair with Alcor
Alpha (α) Centauri	14	39.6	−60	50	0.0 / 1.2			Binary, beautiful pair of stars
Izar (ε Boötis)	14	45.0	+27	04	2.4 / 5.1	2.9	344	Fine pair of yellow and blue stars
Omega[1,2] (ω[1,2]) Scorpii	16	06.0	−20	41	4.0 / 4.3	14.6	145	Optical pair, easy
Epsilon[1] (ε[1]) Lyrae	18	44.3	+39	40	4.7 / 6.2	2.6	346	The Double-Double, quadruple system with ε[2]
Epsilon[2] (ε[2]) Lyrae	18	44.3	+39	40	5.1 / 5.5	2.3	76	Both individual pairs just visible in 80mm telescopes
Theta[1,2] (θ[1,2]) Serpentis	18	56.2	+04	12	4.6 / 5.0	22.4	104	Easy pair, mag 6.7 yellow star 7 arc minutes from θ[2]

Star	RA		Declination		Magnitudes	Separation	PA	Comments
	h	m	°	‘		(arcsec)	°	
Albireo (β Cygni)	19	30.7	+27	58	3.1 / 5.1	34.3	54	Glorious pair, yellow and blue-green
Algedi (α¹,² Capricorni)	20	18.0	−12	32	3.7 / 4.3	6.3	292	Optical pair, easy
Gamma (γ) Delphini	20	46.7	+16	07	5.14 / 4.27	9.2	265	Easy, orange and yellow-white
61 Cygni	21	06.9	+38	45	5.20 / 6.05	31.6	152	Binary, 678 years, both orange
Theta (θ) Indi	21	19.9	−53	27	4.6 / 7.2	7.0	275	Fine object for small telescopes
Delta (δ) Tucanae	22	27.3	−64	58	4.49 / 8.7	7.0	281	Beautiful double, white and reddish

Some Interesting Nebulae, Star Clusters and Galaxies

Brian Jones

Object	RA h	m	Declination °	‘	Remarks
47 Tucanae (in Tucana)	00	24.1	−72	05	Fine globular cluster, easy with naked eye
M31 (in Andromeda)	00	40.7	+41	05	Andromeda Galaxy, visible to unaided eye
Small Magellanic Cloud	00	52.6	−72	49	Satellite galaxy of the Milky Way
NGC 362 (in Tucana)	01	03.3	−70	51	Globular cluster, impressive sight in telescopes
M33 (in Triangulum)	01	31.8	+30	28	Triangulum Spiral Galaxy, quite faint
NGC 869 and NGC 884	02	20.0	+57	08	Sword Handle Double Cluster in Perseus
M34 (in Perseus)	02	42.1	+42	46	Open star cluster near Algol
M45 (in Taurus)	03	47.4	+24	07	Pleiades or Seven Sisters cluster, a fine object
Large Magellanic Cloud	05	23.5	−69	45	Satellite galaxy of the Milky Way
30 Doradus (in Dorado)	05	38.6	−69	06	Star-forming region in Large Magellanic Cloud
M1 (in Taurus)	05	32.3	+22	00	Crab Nebula, near Zeta (ζ) Tauri
M38 (in Auriga)	05	28.6	+35	51	Open star cluster
M42 (in Orion)	05	33.4	−05	24	Orion Nebula
M36 (in Auriga)	05	36.2	+34	08	Open star cluster
M37 (in Auriga)	05	52.3	+32	33	Open star cluster
M35 (in Gemini)	06	06.5	+24	21	Open star cluster near Eta (η) Geminorum
M41 (in Canis Major)	06	46.0	−20	46	Open star cluster to south of Sirius
M44 (in Cancer)	08	38.0	+20	07	Praesepe, visible to naked eye
M81 (in Ursa Major)	09	55.5	+69	04	Bode's Galaxy
M82 (in Ursa Major)	09	55.9	+69	41	Cigar Galaxy or Starburst Galaxy
Carina Nebula (in Carina)	10	45.2	−59	52	NGC 3372, large area of bright and dark nebulosity
M104 (in Virgo)	12	40.0	−11	37	Sombrero Hat Galaxy to south of Porrima
Coal Sack (in Crux)	12	50.0	−62	30	Prominent dark nebula, visible to naked eye
NGC 4755 (in Crux)	12	53.6	−60	22	Jewel Box open cluster, magnificent object
Omega (ω) Centauri	13	23.7	−47	03	Splendid globular in Centaurus, easy with naked eye
M51 (in Canes Venatici)	13	29.9	+47	12	Whirlpool Galaxy
M3 (in Canes Venatici)	13	40.6	+28	34	Bright Globular Cluster

Object	RA		Declination		Remarks
	h	m	°	'	
M4 (in Scorpius)	16	21.5	−26	26	Globular cluster, close to Antares
M12 (in Ophiuchus)	16	47.2	−01	57	Globular cluster
M10 (in Ophiuchus)	16	57.1	−04	06	Globular cluster
M13 (in Hercules)	16	40.0	+36	31	Great Globular Cluster, just visible to naked eye
M92 (in Hercules)	17	16.1	+43	11	Globular cluster
M6 (in Scorpius)	17	36.8	−32	11	Open cluster
M7 (in Scorpius)	17	50.6	−34	48	Bright open cluster
M20 (in Sagittarius)	18	02.3	−23	02	Trifid Nebula
M8 (in Sagittarius)	18	03.6	−24	23	Lagoon Nebula, just visible to naked eye
M16 (in Serpens)	18	18.8	−13	49	Eagle Nebula and star cluster
M17 (in Sagittarius)	18	20.2	−16	11	Omega Nebula
M11 (in Scutum)	18	49.0	−06	19	Wild Duck open star cluster
M57 (in Lyra)	18	52.6	+32	59	Ring Nebula, brightest planetary
M27 (in Vulpecula)	19	58.1	+22	37	Dumbbell Nebula
M29 (in Cygnus)	20	23.9	+38	31	Open cluster
M15 (in Pegasus)	21	28.3	+12	10	Bright globular cluster near Epsilon (ε) Pegasi
M39 (in Cygnus)	21	31.6	+48	25	Open cluster, good with low powers
M52 (in Cassiopeia)	23	24.2	+61	35	Open star cluster near 4 Cassiopeiae

M = Messier Catalogue Number   NGC = New General Catalogue Number

The positions in the sky of each of the objects contained in this list are given on the Monthly Star Charts printed elsewhere in this volume.

Astronomical Organizations

American Association of Variable Star Observers

49 Bay State Road, Cambridge, Massachusetts 02138, USA
www.aavso.org
The AAVSO is an international non-profit organization of variable star observers whose mission is to enable anyone, anywhere, to participate in scientific discovery through variable star astronomy. We accomplish our mission by carrying out the following activities:

- observation and analysis of variable stars
- collecting and archiving observations for worldwide access
- forging strong collaborations between amateur and professional astronomers
- promoting scientific research, education and public outreach using variable star data

American Astronomical Society

1667 K Street NW, Suite 800, Washington, DC 20006, USA
https://aas.org
Established in 1899, the American Astronomical Society (AAS) is the major organization of professional astronomers in North America. The mission of the AAS is to enhance and share humanity's scientific understanding of the universe, which it achieves through publishing, meeting organization, education and outreach, and training and professional development.

Astronomical Society of the Pacific

390 Ashton Avenue, San Francisco, CA 94112, USA
www.astrosociety.org
Formed in 1889, the Astronomical Society of the Pacific (ASP) is a non-profit membership organization which is international in scope. The mission of the ASP is to increase the understanding and appreciation of astronomy through

the engagement of our many constituencies to advance science and science literacy. We invite you to explore our site to learn more about us; to check out our resources and education section for the researcher, the educator, and the backyard enthusiast; to get involved by becoming an ASP member; and to consider supporting our work for the benefit of a science literate world!

Astrospeakers.org
www.astrospeakers.org

A website designed to help astronomical societies and clubs locate astronomy and space lecturers which is also designed to help people find their local astronomical society. It is completely free to register and use and, with over 50 speakers listed, is an excellent place to find lecturers for your astronomical society meetings and events. Speakers and astronomical societies are encouraged to use the online registration to be added to the lists.

British Astronomical Association
Burlington House, Piccadilly, London, W1J 0DU, England
www.britastro.org

The British Astronomical Association is the UK's leading society for amateur astronomers catering for beginners to the most advanced observers who produce scientifically useful observations. Our Observing Sections provide encouragement and advice about observing. We hold meetings around the country and publish a bi-monthly Journal plus an annual Handbook. For more details, including how to join the BAA or to contact us, please visit our website.

British Interplanetary Society
Arthur C Clarke House, 27/29 South Lambeth Road, London, SW8 1SZ, England
www.bis-space.com

The British Interplanetary Society is the world's longest-established space advocacy organisation, founded in 1933 by the pioneers of British astronautics. It is the first organisation in the world still in existence to design spaceships. Early members included Sir Arthur C Clarke and Sir Patrick Moore. The Society has created many original concepts, from a 1938 lunar lander and space suit designs, to geostationary orbits, space stations and the first engineering study of a starship, Project Daedalus. Today the BIS has a worldwide membership and welcomes all with an interest in Space, including enthusiasts, students, academics and professionals.

Canadian Astronomical Society
Société Canadienne D'astronomie (CASCA)

100 Viaduct Avenue West, Victoria, British Columbia, V9E 1J3, Canada
www.casca.ca

CASCA is the national organization of professional astronomers in Canada. It seeks to promote and advance knowledge of the universe through research and education.   Founded in 1979, members include university professors, observatory scientists, postdoctoral fellows, graduate students, instrumentalists, and public outreach specialists.

Royal Astronomical Society of Canada

203-4920 Dundas St W, Etobicoke, Toronto, ON M9A 1B7, Canada
www.rasc.ca

Bringing together over 5,000 enthusiastic amateurs, educators and professionals RASC is a national, non-profit, charitable organization devoted to the advancement of astronomy and related sciences and is Canada's leading astronomy organization. Membership is open to everyone with an interest in astronomy. You may join through any one of our 29 RASC centres, located across Canada and all of which offer local programs. The majority of our events are free and open to the public.

Federation of Astronomical Societies

The Secretary, 147 Queen Street, SWINTON, Mexborough, S64 8NG
www.fedastro.org.uk

The Federation of Astronomical Societies (FAS) is an umbrella group for astronomical societies in the UK. It promotes cooperation, knowledge and information sharing and encourages best practice. The FAS aims to be a body of societies united in their attempts to help each other find the best ways of working for their common cause of creating a fully successful astronomical society. In this way it endeavours to be a true federation, rather than some remote central organization disseminating information only from its own limited experience. The FAS also provides a competitive Public Liability Insurance scheme for its members.

International Dark-Sky Association
darksky.org

The International Dark-Sky Association (IDA) is the recognized authority on light pollution and the leading organization combating light pollution worldwide. The IDA works to protect the night skies for present and future generations, our public outreach efforts providing solutions, quality education and programs that inform audiences across the United States of America and throughout the world. At the local level, our mission is furthered through the work of our U.S. and international chapters representing five continents.
The goals of the IDA are:

- Advocate for the protection of the night sky
- Educate the public and policymakers about night sky conservation
- Promote environmentally responsible outdoor lighting
- Empower the public with the tools and resources to help bring back the night

Royal Astronomical Society of New Zealand
PO Box 3181, Wellington, New Zealand
www.rasnz.org.nz

Founded in 1920, the object of The Royal Astronomical Society of New Zealand is the promotion and extension of knowledge of astronomy and related branches of science. It encourages interest in astronomy and is an association of observers and others for mutual help and advancement of science. Membership is open to all interested in astronomy. The RASNZ has about 180 individual members including both professional and amateur astronomers and many of the astronomical research and observing programmes carried out in New Zealand involve collaboration between the two. In addition the society has a number of groups or sections which cater for people who have interests in particular areas of astronomy.

Astronomical Society of Southern Africa
Astronomical Society of Southern Africa, c/o SAAO, PO Box 9, Observatory, 7935, South Africa
assa.saao.ac.za

Formed in 1922, The Astronomical Society of Southern Africa comprises both amateur and professional astronomers. Membership is open to all interested persons. Regional Centres host regular meetings and conduct public outreach events, whilst national Sections coordinate special interest groups and observing programmes. The Society administers two Scholarships, and hosts occasional Symposia where papers are presented. For more details, or to contact us, please visit our website.

Royal Astronomical Society
Burlington House, Piccadilly, London, W1J 0BQ, England
www.ras.org.uk
The Royal Astronomical Society, with around 4,000 members, is the leading UK body representing astronomy, space science and geophysics, with a membership including professional researchers, advanced amateur astronomers, historians of science, teachers, science writers, public engagement specialists and others.

Society for Popular Astronomy
Secretary: Guy Fennimore, 36 Fairway, Keyworth, Nottingham, NG12 5DU
www.popastro.com
The Society for Popular Astronomy is a national society that aims to present astronomy in a less technical manner. The bi-monthly society magazine Popular Astronomy is issued free to all members.

Our Contributors

John C. Barentine is the Director of Public Policy for the International Dark-Sky Association (IDA). He earned a Ph.D. in astronomy from the University of Texas at Austin, and has held staff positions at the National Solar Observatory, Apache Point Observatory, and the Sloan Digital Sky Survey. Throughout his career he has been involved in education and outreach efforts to help increase the public understanding of astronomy and science in general. In addition to his work for the IDA, John is a founding member of the University of Utah Consortium for Dark Sky Studies, serves on light pollution committees of both the American Astronomical Society and the International Astronomical Union, and is a Fellow of the Royal Astronomical Society. John is the author of two books on the history of astronomy, *The Lost Constellations* and *Uncharted Constellations*.

Martin Braddock is a professional scientist and project leader in the field of drug discovery and development with 34 years' experience of working in academic institutes and large corporate organizations. He holds a BSc in Biochemistry and a PhD in Radiation Biology and is a former Royal Society University Research Fellow at the University of Oxford. He was elected a Fellow of the Royal Society of Biology in 2010, and in 2012 received an Alumnus Achievement Award for distinction in science from the University of Salford. Martin has published over 170 peer-reviewed scientific papers, filed 9 patents, edited 2 books for the Royal Society of Chemistry and serves as a proposal evaluator for multiple international research agencies. Martin holds further qualifications from the University of Central Lancashire and Open University. He is a member of the Mansfield and Sutton Astronomical Society and was elected a Fellow of the Royal Astronomical Society in May 2015. An ambassador for science, technology, engineering and mathematics (STEM), Martin seeks to inspire the next generation of young scientists to aim high and be the best they can be. To find out more about him visit science4u.co.uk

Jane Callaghan was, at the time of writing her article on Alfred Fowler, a Parish Councillor in the Pennine mill village of Wilsden. One day she googled 'Born in Wilsden' and up popped Alfred Fowler. Curiosity piqued, Jane researched him and crowd funded money for a blue plaque which was erected in time for the 150th anniversary of his birth.

Paul Fellows is chairman of the Cambridge Astronomical Association and co-presents the weekly Cambridge University Institute of Astronomy public open nights from October through to March. Having built his first telescope at the age of 14, he has always kept a keen eye on developments in astronomy while pursuing a career in software and electronics. Twice winner of the Queens Award for Technology, Paul holds Degrees in Natural Sciences and in Computer Science from Cambridge and is an elected Fellow of the Institute of Engineering and Technology and of the Royal Astronomical Society.

Neil Haggath has a degree in astrophysics from Leeds University and has been a Fellow of the Royal Astronomical Society since 1993. A member of Cleveland and Darlington Astronomical Society for 40 years, he has served on its committee for 32 years. Neil is an avid umbraphile, clocking up six total eclipse expeditions so far, to locations as far flung as Australia and Hawai'i. Four of them were successful, the most recent being in Jackson, Wyoming on 21 August 2017. In 2012, he may have set a somewhat unenviable record among British astronomers - for the greatest distance travelled (6,000 miles to Thailand) to NOT see the transit of Venus. He saw nothing on the day … and got very wet!

Jan Hardy first became interested in astronomy when she was seven years old after watching the Apollo 11 moon mission live on television. She went straight out and bought *The Observer's Book of the Moon* with her pocket money. This interest neatly fitted in with her other great interest - science fiction. Both have had a great influence on Jan throughout her life, albeit astronomy mostly in the background as a solitary pastime until joining Bradford Astronomical Society in January 2000 (an end-of-Millennium New Year's Resolution). Although she is interested in all aspects of astronomy, in recent years Jan has become particularly interested in cosmology, and enjoys regular discussions on the subject with fellow amateur astronomers.

Dr. David M. Harland gained his BSc in astronomy in 1977, lectured in computer science, worked in industry, and managed academic research. In 1995 he 'retired' in order to write on space themes.

David Harper, FRAS has had a varied career which includes teaching mathematics, astronomy and computing at Queen Mary University of London, astronomical software development at the Royal Greenwich Observatory, bioinformatics support at the Wellcome Trust Sanger Institute, and a research interest in the dynamics of planetary satellites, which began during his Ph.D. at Liverpool University in the 1980s and continues in an occasional collaboration with colleagues in China. He is married to fellow contributor Lynne Stockman.

Dr. Pauline Harris is from the tribes Rongomaiwahine, Ngāti Rakaipaka and Ngāti Kahungunu. She is a Senior Lecturer for the Centre for Science and Society at Victoria University of Wellington (VUW). Dr Harris has a background in physics and completed her PhD and Masters at the University of Canterbury, where her research focussed on gamma ray bursts, high-energy neutrino production and inflationary cosmology. She has also lectured in Physics at VUW and now focuses on mātauranga Māori associated with Māori astronomy and traditional Māori calendars called maramataka. Pauline is extensively involved in Māori communities and is an active member of a number of trust boards. She is currently the Chairperson of the Society for Māori Astronomy Research and Traditions (SMART), where she is dedicated to the collation and the revitalisation of Māori astronomical star lore and maramataka. Her research interests and involvement with these communities and boards have led to the development of a number of large research projects that have been funded by the likes of the Marsden Fund, Ministry of Business Innovation and Employment, Te Puni Kōkiri, The Māori Language Commission and the United Nations Education Science and Cultural Organisation (UNESCO).

Tracie Heywood is an amateur astronomer from Leek in Staffordshire and is one of the UK's leading variable star observers, using binoculars to monitor the brightness changes of several hundred variable stars. Tracie currently writes a monthly column about variable stars for *Astronomy Now* magazine. She has previously been the Eclipsing Binary coordinator for the Variable Star Section of the British Astronomical Association and the Director of the Variable Star Section of the Society for Popular Astronomy.

Rod Hine was aged around ten when he was given a copy of *The Boys Book of Space* by Patrick Moore. Already interested in anything to do with science and engineering he devoured the book from cover to cover. The launch of Sputnik I shortly afterwards clinched his interest in physics and space travel. He took physics, chemistry and mathematics at A-level and then studied Natural Sciences at Churchill College, Cambridge. He later switched to Electrical Sciences and subsequently joined Marconi at Chelmsford working on satellite communications in the UK, Middle East and Africa. This led to work in meteorological communications in Nairobi, Kenya and later a teaching post at the Kenya Polytechnic. There he met and married a Yorkshire lass and moved back to the UK in 1976. Since then he has had a variety of jobs in electronics and industrial controls, and until recently was lecturing part-time at the University of Bradford. Rod got fully back into astronomy in around 1992 when his wife bought him an astronomy book, at which time he joined Bradford Astronomical Society. He is currently working part-time at Leeds University providing engineering support for a project to convert redundant satellite dishes into radio telescopes in developing countries.

Brian Jones hails from Bradford in the West Riding of Yorkshire and was a founder member of the Bradford Astronomical Society. He developed a fascination with astronomy at the age of five when he first saw the stars through a pair of binoculars, although he spent the first part of his working life developing a career in mechanical engineering. However, his true passion lay in the stars and his interest in astronomy took him into the realms of writing sky guides for local newspapers, appearing on local radio and television, teaching astronomy and space in schools and, in 1985, leaving engineering to become a full time astronomy and space writer. His books have covered a range of astronomy and space-related topics for both children and adults and his journalistic work includes writing articles and book reviews for several astronomy magazines as well as for many general interest magazines, newspapers and periodicals. His passion for bringing an appreciation of the universe to his readers is reflected in his writing.

You can follow Brian on Twitter via @StarsBrian and check out the sky by visiting his blog at **starlight-nights.co.uk** from where you can also access his Facebook group Starlight Nights.

Carolyn Kennett lives in the far south west of Cornwall. She likes to write, although you will often find her walking the Tinner's Way and coastal pathways found in the local countryside. As well as researching local astronomy history, she has a passion for archaeoastronomy and eighteenth- and nineteenth-century maritime astronomy. She currently co-edits *Bulletin*, the magazine of the Society for the History of Astronomy. Further details of the SHA can be found at **societyforthehistoryofastronomy.com**

John McCue graduated in astronomy from the University of St Andrews and began teaching. He gained a Ph.D. from Teesside University studying the unusual rotation of Venus. In 1979 he and his colleague John Nichol founded the Cleveland and Darlington Astronomical Society, which then worked in partnership with the local authority to build the Wynyard Planetarium and Observatory in Stockton-on-Tees. John is currently double star advisor for the British Astronomical Association.

Mary McIntyre is an amateur astronomer, astrophotographer and astronomy sketcher based in Oxfordshire, UK. She has had a life-long interest in astronomy and loves sharing that with others. She is a writer and communicator and has contributed several articles to *Sky at Night* magazine and made numerous radio appearances including being a co-presenter of the monthly radio show Comet Watch. She also gives regular astronomy and photography talks to people of all ages, including presentations to astronomy societies, camera clubs, schools and Beavers/Cubs/Scouts groups. She is passionate about promoting science, technology, engineering and mathematics (STEM) and astronomy to women and girls and runs the UK Women in Astronomy Network, which connects, celebrates and promotes women with a passion for astronomy. By providing role models from both professional and the keen amateur, women and young girls are encouraged and supported to explore all aspects of astronomy and astrophotography and given the confidence to pursue a career in the field. To find out more about Mary visit www.marymcintyreastronomy.co.uk

Rangi Matamua of Tūhoe holds the position of Associate Dean Postgraduate in the Faculty of Māori and Indigenous Studies at Waikato University. Professor Matamua has a background in Māori language, Māori broadcasting and Māori customs and traditions and is acknowledged as a leading expert in the field of Māori astronomy. He has delivered many Māori astronomy keynote addresses and public lectures, both nationally and internationally, and in 2017 he launched his best-selling first book *Matariki – the Star of the Year* in both English and te reo Māori. A graduate of Te Panekiretanga o te reo Māori and a member of the Society for Māori Astronomy Research and Traditions (SMART), Professor Matamua is focused on revitalising Māori astronomy.

Peter Meadows is an Assistant Director of the BAA Solar Section and the solar sub-editor for *The Astronomer* magazine. He became interested in solar observing in the late 1970s and uses modest equipment: an 80mm refractor for white light observing, a Coronado PST for hydrogen alpha observing and imaging, an ETX 105 for white light imaging and a VLF receiver for flare detection. The freeware Helio software programs, available from www.petermeadows.com, were created to analyse his disk drawings. His white light observations also contribute to the International Sunspot Number.

Peter also enjoys observing other astronomical objects such as meteors, noctilucent clouds and occasionally the planets. He has also used remote telescopes for the imaging and measurement of asteroids, comets and variable stars. In 2017 he received the BAA Steavenson Award for contributing to observational astronomy.

Professionally he is a research scientist specialising on Earth observation using European Space Agency Synthetic Aperture Radar satellites.

Neil Norman, FRAS first became fascinated with the night sky when he was five years of age and saw Patrick Moore on the television for the first time. It was the Sky at Night programme, broadcast in March 1986 and dedicated to the Giotto probe reaching Halley's Comet, which was to ignite his passion for these icy interlopers. As the years passed, he began writing astronomy articles for local news magazines before moving into internet radio where he initially guested on the Astronomyfm show 'Under British Skies', before becoming a co-host for a short time. In 2013 he created Comet Watch, a Facebook group dedicated to comets of the past, present and future. His involvement with Astronomyfm led to the creation of the monthly radio show Comet Watch, which is now in its fourth year. Neil lives in Suffolk with his partner and three children. Perhaps rather fittingly, given Neil's interest in asteroids, he has one named in his honour, this being the main belt asteroid 314650 Neilnorman, discovered in July 2006 by English amateur astronomer Matt Dawson.

Peter Rea has had a keen interest in lunar and planetary exploration since the early 1960s and frequently lectures on the subject. He helped found the Cleethorpes and District Astronomical Society in 1969. In April of 1972 he was at the Kennedy Space Centre in Florida to see the launch of Apollo 16 to the moon and in October 1997 was at the southern end of Cape Canaveral to see the launch of Cassini to Saturn. He would still like to see a total solar eclipse as the expedition he was on to see the 1973 eclipse in Mali had vehicle trouble, and the meteorologists decided he was not going to see the 1999 eclipse from Devon. In 2019 Peter was elected a Fellow of the Royal Astronomical Society. He lives in Lincolnshire with his wife Anne and has a daughter who resides in New Zealand.

Richard Sanderson retired in 2018 after a 19-year tenure as curator of physical science at the Springfield Science Museum in Massachusetts, where he managed the Seymour Planetarium, the oldest operating projection planetarium in the United States. Richard wrote a newspaper astronomy column for many years and co-authored the 2006 book *Illustrated Timeline of the Universe*. He has served as president of the Springfield Stars Club and helps organize the Connecticut River Valley Astronomers' Conjunction, an annual convention of astronomy enthusiasts. His passion for antiquarian astronomy books and memorabilia has flourished for more than 40 years and led to his recent creation of the "Vintage Astronomy Books" Facebook group, located at **www.facebook.com/groups/ vintageastronomy**

Lynne Marie Stockman holds degrees in mathematics from Whitman College, the University of Washington and the University of London, and has studied astronomy at both undergraduate and postgraduate levels. She is a native of North Idaho but has lived in Britain for the past 29 years. Lynne was an early pioneer of the World Wide Web: with her husband David Harper, she created the web site obliquity.com in 1998 to share their interest in astronomy, computing, family history and cats.

Hēmi Whaanga is Ngāti Kahungunu, Ngāti Tahu, Ngāti Mamoe, and Waitaha. He is Associate Dean Research in the Faculty of Māori and Indigenous Studies at the University of Waikato, with research interests focussing on Mātauranga Māori including traditional knowledge, Māori astronomy, traditional ecological knowledge, indigenous ethics, digitisation of indigenous knowledge, indigenous taxonomy, and naming. His work also focuses on linguistics, including discourse analysis, te reo Māori, applied linguistics and language curriculum.

Lionel Wilson has been a planetary science researcher since starting work on his Ph.D. project at the University of London in 1965 during the run-up to the Apollo landings. After a two-year post-doctoral position in London he joined Lancaster University in 1970, where he has guided more than 30 graduate students to obtain their Ph.D.s. He is now Emeritus Professor of Earth and Planetary Sciences at Lancaster. He also holds Visiting Professorships at Brown University (since 1978) and at the University of Hawai'i (since 1984), and normally spends three months of each year in the USA. He has been both a principle investigator and an associate investigator on various NASA planetary missions and specializes in the theoretical modelling of the fluid-mechanical processes associated with explosive and effusive volcanic eruptions on the silicate planets.

Gary Yule developed an interest in astronomy at the age of seven when his father woke him in the middle of the night to observe Halley's Comet. Since then he has become a keen amateur astronomer and his journey has taken him down many avenues, ranging from the history of astronomy to astrophotography. Gary has particular interest in antique telescopes and the stories behind them, and is the Chairman and Curator of Instruments for Salford Astronomical Society where he dedicates most of his time. He also writes monthly sky notes for a local publication and heads up various astronomy and astrophotography pages on social media, in addition to which he buys and sells telescopes and mounts, many of which are antique.

Glossary

Brian Jones and David Harper

Airburst
The violent 'explosion' and resulting energy shockwave of a small *asteroid* or *meteorite* which has entered the Earth's atmosphere, and which occurs before the object reaches the ground. The Tunguska event of 1908 is believed to have been an asteroid airburst.

Albedo
The fraction of incoming light reflected back into space by a body such as a *planet*, *comet* or *satellite*. Objects with high albedos (near 1) are very bright, while those with low albedos (near 0) are relatively dark.

Altitude
The altitude of a star or other object is its angular distance above the horizon. For example, if a star is located at the *zenith*, or overhead point, its altitude is 90° and if it is on the horizon, its altitude is 0°.

Angular Distance
The angular distance between two objects on the sky is the angle subtended between the directions to the two objects, either at the centre of the Earth (geocentric angular distance) or the observer's eye (apparent angular distance). It is most commonly expressed in degrees, or for smaller angular distances, minutes of arc or seconds of arc.

Antoniadi Scale
A scale of *seeing* conditions named after astronomer Eugène Michel Antoniadi who devised it during the early 1900s. It assesses the weather and seeing conditions under which astronomical observations are carried out. The Antoniadi scale has five gradations, these being: (I) perfect seeing with no quivering; (II) good seeing, some slight undulations with frequent steady moments; (III) moderate seeing, about equal steady and turbulent moments; (IV) poor seeing, with constant undulations making sketching difficult; and (V) very bad seeing, with turbulence scarcely allowing a sketch to be made.

Aphelion
This is the point at which an object, such as a *planet*, *comet* or *asteroid* travelling in an elliptical *orbit*, is at its maximum distance from the Sun.

Apogee
This is the point in its *orbit* around the Earth at which an object is at its furthest from the Earth.

Apparition
The period during which a planet is visible, usually starting at *conjunction* with the Sun, running through *opposition* (for a superior planet) or *greatest elongation* (for Mercury or Venus), and ending with the next conjunction with the Sun.

Appulse
The close approach, as seen from Earth, between two planets, or a planet and a star, or the Moon and a star or planet. Also known as a *conjunction*.

Asterism
An asterism is grouping or collection of stars often (but not always) located within a *constellation* that forms an apparent and distinctive pattern in its own right. Well known examples include the Plough (in Ursa Major); the False Cross (formed from stars in Carina and Vela); and the Summer Triangle, which is formed from the bright stars Vega (in Lyra), Deneb (in Cygnus) and Altair (in Aquila).

Asteroid
Another name for a *minor planet*.

Astronomical Unit
Often used to measure distances within our *Solar System*, the astronomical unit (AU) is a unit of measurement equal to the average distance between Earth and the Sun, or around 150 million kilometres (93 million miles).

Autumnal Equinox
The autumnal equinox is the point at which the apparent path of the Sun, moving from north to south, crosses the *celestial equator*. In the Earth's northern hemisphere this marks the start of autumn, whilst in the southern hemisphere it is the start of spring.

Averted Vision
Averted vision is a useful technique for observing faint objects which involves looking slightly to one side of the object under observation and, by doing so, allowing the light emitted by the object to fall on the part of the retina that is more sensitive at low light levels. Although you are not looking directly at the object, it is surprising how much more detail comes into view. This technique is also useful when observing double stars which have components of greatly contrasting brightness. Although direct vision may not reveal the glow of a faint companion star in the glare of a much brighter primary, averted vision may well bring the fainter star into view.

Azimuth
The azimuth of a star or other object is its angular position measured round the *horizon* from north (azimuth 0°) through east (azimuth 90°), south (azimuth 180°) and west (azimuth 270°). The azimuth and *altitude*, taken together, define the position of the object referred to the observer's *local horizon*.

Barycentre
The barycentre is the centre of mass of two or more bodies that are orbiting each other (such as a planet and satellite or two components of a *binary star* system) and which is therefore the point around which they both *orbit*.

Binary Star
See Double Star

Black Hole
A region of space around a very compact and extremely massive collapsed star within which the gravitational field is so intense that not even light can escape.

Caldwell Catalogue

This is a catalogue of 109 star clusters, nebulae, and galaxies compiled by Patrick Moore to complement the **Messier Catalogue**. Intended for use as an observing guide by amateur astronomers it includes a number of bright **deep sky objects** that did not find their way into the Messier Catalogue, which was originally compiled as a list of known objects that might be confused with comets. Moore used his other surname (Caldwell) to name the list and the objects within it (the first letter of 'Moore' having been used for the Messier Catalogue) and entries in the Caldwell Catalogue are designated with a 'C' followed by the catalogue number (1 to 109).

Amongst the 109 objects in the Caldwell Catalogue are the Sword Handle Double Cluster NGC 869 and NGC 884 (C14) in Perseus; supernova remnant(s) the East Veil Nebula and West Veil Nebula (C33 and C34) in Cygnus; the Hyades open star cluster (C41) in Taurus; and Hubble's Variable Nebula (C46) in Monoceros. Unlike the Messier Catalogue, which was compiled from observations made by Charles Messier from Paris, the Caldwell Catalogue contains deep sky objects visible from the southern hemisphere, such as the Centaurus A galaxy (C77) and globular cluster Omega Centauri (C80) in Centaurus; the Jewel Box open star cluster (C94) in Crux and the globular cluster 47 Tucanae (C106) in Tucana.

Although few of the objects detailed elsewhere in the Yearbook of Astronomy carry a Caldwell Catalogue reference, it was felt that an entry should appear in the Glossary as the catalogue is nonetheless an important guide for the backyard astronomer.

Celestial Equator

The celestial equator is a projection of the Earth's **equator** onto the **celestial sphere**, equidistant from the **celestial poles** and dividing the celestial sphere into two hemispheres.

Celestial Poles

The north (and south) celestial poles are points on the **celestial sphere** directly above the north and south terrestrial poles around which the celestial sphere appears to rotate and through which extensions of the Earth's axis of rotation would pass.

The north celestial pole, the position of which is at marked at present by the relatively bright star Polaris, lies in the constellation Ursa Minor (the Little Bear) and would be seen directly overhead when viewed from the North Pole. There is no particularly bright star marking the position of the south celestial pole, which lies in the tiny **constellation** Octans (the Octant) and which would be situated directly overhead when seen from the South Pole. The north celestial pole lies in the direction of north when viewed from elsewhere on the Earth's surface and the south celestial pole lies in the direction of south when viewed from other locations.

Celestial Sphere

The imaginary sphere surrounding the Earth on which the stars appear to lie.

Circumpolar Star

A circumpolar star is a star which never sets from a given **latitude**. When viewing the sky from either the North or South Pole, all stars will be circumpolar, although no stars are circumpolar when viewed from the equator.

Comet

A comet is an object comprised of a mixture of gas, dust and ice which travels around the Sun in an orbit that can often be very eccentric.

Conjunction

This is the position at which two objects are lined up with each other (or nearly so) as seen from Earth. Superior conjunction occurs when a planet is at the opposite side of the Sun as seen from Earth and inferior conjunction when a planet lies between the Sun and Earth.

Constellation

A constellation is an arbitrary grouping of stars forming a pattern or imaginary picture on the celestial sphere. Many of these have traditional names and date back to ancient Greece or even earlier and are associated with the folklore and mythology of the time. There are also some of what may be described as 'modern' constellations, devised comparatively recently by astronomers during the last few centuries. There are 88 official constellations which together cover the entire sky, each one of which refers to and delineates that particular region of the *celestial sphere*, the result being that every celestial object is described as being within one particular constellation or another.

Dark Nebula

See Nebula.

Declination

This is the *angular distance* between a celestial object and the celestial equator. Declination is expressed in degrees, minutes and seconds either north (N) or south (S) of the *celestial equator*.

Deep Sky Object

Deep sky objects are objects (other than individual stars) which lie beyond the confines of our *Solar System*. They may be either galactic or extra-galactic and include such things as *star clusters*, *nebulae* and *galaxies*.

Direct Motion

A planet is in direct (or prograde) motion when its *right ascension* or ecliptic *longitude* is increasing with the passing of time. This means that it is moving eastwards with respect to the background stars.

Double Stars

Double stars are two stars which appear to be close together in space. Although some double stars (known as *optical* doubles) are made up of two stars that only happen to lie in the same line of sight as seen from Earth and are nothing more than chance alignments, most are comprised of stars that are gravitationally linked and orbit each other, forming a genuine double-star system (also known as a *binary* star).

Eclipse

An eclipse is the obscuration of one celestial object by another, such as the Sun by the Moon during a solar eclipse or one component of an eclipsing *binary star* by the companion star.

- A **solar eclipse** occurs when the Moon passes directly between the Earth and the Sun. There are three types of solar eclipse. A total solar eclipse takes place when the Moon completely obscures the Sun, during which event the Sun's corona, or outer atmosphere, is revealed; a partial solar eclipse occurs when the lining up of the Earth, Moon and Sun is not exact and the Moon covers only a part of the Sun; an annular solar eclipse takes place when the Moon is at or near its farthest from Earth, at which time the lunar disc appears smaller and does not completely cover the solar disc, the Sun's visible outer edges forming a 'ring of light' or 'annulus' around the Moon. Some eclipses which begin as annular may become total along part of their path; these are known as hybrid eclipses, and are quite rare.
- A **lunar eclipse** occurs when the Earth passes between the Sun and the Moon, and the Earth's shadow is thrown onto the lunar surface. There are three types of lunar eclipse. A total lunar eclipse takes place when the Moon passes completely through the *umbra* of the Earth's shadow, during which process the Moon will gradually darken and take on a reddish / rusty hue; a partial

lunar eclipse occurs when the Moon passes through the *penumbra* of the Earth's shadow and only part of it enters the umbra; a penumbral lunar eclipse takes place when the Moon only enters the penumbra of the Earth's shadow without touching or entering the umbra.

Ecliptic
As the Earth orbits the Sun, its position against the background stars changes slightly from day to day, the overall effect of this being that the Sun appears to travel completely around the *celestial sphere* over the course of a year. The apparent path of the Sun is known as the ecliptic and is superimposed against the band of *constellations* we call the *Zodiac* through which the Sun appears to move.

Ellipse
The closed, oval-shaped form obtained by cutting through a cone at an angle to the main axis of the cone. The orbits of the planets around the Sun are all elliptical.

Elongation (and Greatest Elongation)
In its most general sense, elongation refers to the angular separation between two celestial objects as seen from a third object. It is most often used to refer to the *angular distance* between the Sun and a planet or the Moon, as seen from Earth.

The greatest elongation of Mercury or Venus is the maximum angular distance between the planet and the Sun as seen from Earth, during a particular *apparition*.

Emission Nebula
See Nebula.

Ephemeris (plural: Ephemerides)
Table showing the predicted positions of celestial objects such as comets or planets.

Equator
The equator of a planet or other spheroidal celestial body is the great circle on the surface of the body whose latitude is zero, as defined by the axis of rotation. The *celestial equator* is the projection of the plane of the Earth's equator onto the sky.

Equinox
The equinoxes are the two points at which the ecliptic crosses the *celestial equator* (see also *Autumnal Equinox* and *Vernal Equinox*). The term is also used to denote the dates on which the Sun passes these points on the *ecliptic*.

Exoplanet
An exoplanet (or extrasolar planet) is a planet orbiting a star outside of our *Solar System*.

First Point of Aries
The First Point of Aries is the point on the *celestial sphere* at which the *ecliptic* crosses the *celestial equator* from south to north. The Sun is at the First Point of Aries on the *vernal equinox*. It is the zero point from which *Right Ascension* and ecliptic *longitude* are measured. At present, the First Point of Aries is actually in the constellation of Pisces. Several centuries from now, *precession* will eventually carry it into Aquarius.

Galaxy
A galaxy is a vast collection of stars, gas and dust bound together by gravity and measuring many thousands of light years across. Galaxies occur in a wide variety of shapes and sizes including

spiral, elliptical and irregular and most are so far away that their light has taken many millions of years to reach us. Our *Solar System* is situated in the Milky Way Galaxy, a spiral galaxy containing several billion stars. Located within the *Local Group of Galaxies*, the *Milky Way* Galaxy is often referred to simply as the Galaxy.

Horizon

The horizon is a great circle that is theoretically defined by a zenith distance of 90 degrees. In practice, the observer's *local horizon* will differ from this.

Index Catalogue (IC)

References such as that for IC 2391 (in Vela) and IC 2602 (in Carina) are derived from their numbers in the Index Catalogue (IC), published in 1895 as the first of two supplements (the second was published in 1908) to his *New General Catalogue* of Nebulae and Clusters of Stars (NGC) by the Danish astronomer John Louis Emil Dreyer (1852–1926). Between them, the two Index Catalogues contained details of an additional 5,386 objects.

Inferior Planet

An inferior planet is a planet that travels around the Sun inside the *orbit* of the Earth.

International Astronomical Union (IAU)

Formed in 1919 and based at the Institut d'Astrophysique de Paris, this is the main coordinating body of world astronomy. Its main function is to promote, through international cooperation, all aspects of the science of astronomy. It is also the only authority responsible for the naming of celestial objects and the features on their surfaces.

Latitude

The latitude of the Sun, Moon or planet is its angular distance above or below the *ecliptic*. Note that the *angular distance* of a celestial body north or south of the *celestial equator* is called *declination*, and not latitude.

The latitude of a point on the Earth's surface is its angular distance north or south of the *equator*.

Light Year

To express distances to the stars and other galaxies in miles would involve numbers so huge that they would be unwieldy. Astronomers therefore use the term 'light year' as a unit of distance. A light year is the distance that a beam of light, travelling at around 300,000 km (186,000 miles) per second, would travel in a year and is equivalent to just under 10 trillion km (6 trillion miles).

Local Group of Galaxies

This is a gravitationally-bound collection of galaxies which contains over 50 individual members, one of which is our own Milky Way Galaxy. Other members include the Large Magellanic Cloud, the Small Magellanic Cloud, the Andromeda Galaxy (M31), the Triangulum Spiral Galaxy (M33) and many others.

Galaxies are usually found in groups or clusters. Apart from our own Local Group, many other groups of galaxies are known, typically containing anywhere up to 50 individual members. Even larger than the groups are clusters of galaxies which can contain hundreds or even thousands of individual galaxies. Groups and clusters of galaxies are found throughout the universe.

Local Horizon

The horizon seen by an observer on land or at sea differs from the ideal theoretical horizon, defined as 90 degrees from the *zenith*, due to several factors. This can affect astronomical observations.

On land, distant features such as mountains may delay the appearance of the rising Sun, Moon or stars by minutes or even hours compared to rising times tabulated in almanacs. At sea, altitudes measured relative to the sea horizon are affected by the observer's height above sea level. At a height of 30 metres above sea level (an aircraft carrier deck, for example), this 'dip' of the sea horizon is 10 arc-minutes, and the **altitude** of a star observed using a nautical sextant must have this amount subtracted before it can be used to determine position at sea. The effect may seem small, but 1 arc-minute of observed altitude corresponds to one nautical mile, so ignoring the 10 arc-minute dip correction would lead to an error of 10 nautical miles in the position of the ship.

Local Hour Angle
The local hour angle of a star or other celestial object is the difference between the local **sidereal time** and the object's **right ascension**. At upper **transit**, an object's local hour angle is zero. Before transit, the local hour angle is negative, whilst after transit, it is positive.

Longitude
The longitude of the Sun, Moon or planet is its angular position, measured along the **ecliptic** from the **First Point of Aries**.

The longitude of a point on the Earth's surface is its **angular distance** east or west of the **prime meridian** through Greenwich. By convention, terrestrial longitude is positive east of Greenwich and negative west of Greenwich.

Lucida
The brightest **star** in a **constellation**.

Lunar
Of or appertaining to the Moon.

Lunar Eclipse
See Eclipse.

Magnitude (apparent)
The magnitude of a star is purely and simply a measurement of its brightness. In around 150BC the Greek astronomer Hipparchus divided the stars up into six classes of brightness, the most prominent stars being ranked as first class and the faintest as sixth. This system classifies the stars and other celestial objects according to how bright they actually appear to the observer. In 1856 the English astronomer Norman Robert Pogson refined the system devised by Hipparchus by classing a first-magnitude star as being 100 times as bright as one of sixth-magnitude, giving a difference between successive magnitudes of $\sqrt[5]{100}$ or 2.512. In other words, a star of magnitude 1.00 is 2.512 times as bright as one of magnitude 2.00, 6.31 (2.512 × 2.512) times as bright as a star of magnitude 3.00 and so on. The same basic system is used today, although modern telescopes enable us to determine values to within 0.01 of a magnitude or better. Negative values are used for the brightest objects including the Sun (−26.8), the Full Moon (−12.9), Venus (−4.7 at its brightest) and Sirius (−1.46). Generally speaking, the faintest objects that can be seen with the naked eye under good viewing conditions are around sixth-magnitude, with binoculars allowing you to see stars and other objects down to around ninth-magnitude.

Magnitude (absolute)
The **apparent magnitude** of a **star** is not a reliable measure of the actual (intrinsic) luminosity of that star. For example, Deneb and Rigil Kentaurus are both first-magnitude stars as seen from Earth, but Deneb is 130,000 times more luminous than Rigil Kentaurus. The latter star appears

brighter only because it is much closer to us. Astronomers correct for this distance effect by defining the absolute magnitude of a star as the *apparent magnitude* it would have if it were located at a standard distance of 10 *parsecs* (32.6 *light years*). This enables astronomers to compare the intrinsic luminosities of stars directly. The absolute magnitude of the Sun is +4.87 whilst Rigil Kentaurus and Deneb have absolute magnitudes of +4.43 and −8.38 respectively.

Meridian
This is a great circle crossing the *celestial sphere* and which passes through both *celestial poles* and the *zenith*.

Messier Catalogue and References
References such as that for Messier 1 (M1) in Taurus, Messier 31 (M31) in Andromeda and Messier 57 (M57) in Lyra relate to a range of deep sky objects derived from the *Catalogue des Nébuleuses et des Amas d'Étoiles* (Catalogue of Nebulae and Star Clusters) drawn up by the French astronomer Charles Messier during the latter part of the eighteenth century.

Meteor
This is a streak of light in the sky seen as the result of the destruction through atmospheric friction of a *meteoroid* in the Earth's atmosphere.

Meteorite
A meteorite is a *meteoroid* which is sufficiently large to at least partially survive the fall through Earth's atmosphere.

Meteoroid
This is a term applied to particles of interplanetary meteoritic debris.

Milky Way
This is the name given to the faint pearly band of light that we sometimes see crossing the sky and which is formed from the collective glow of the combined light from the thousands of stars that lie along the main plane of our Galaxy as seen from Earth. The vast majority of these stars are too faint to be seen individually without some form of optical aid. However, provided the sky is really dark and clear, the Milky Way itself is easily visible to the unaided eye, and any form of optical aid will show that it is indeed made up of many thousands of individual stars. Our *Solar System* lies within the main plane of the Milky Way Galaxy and is located inside one of its spiral arms. The Milky Way is actually our view of the Galaxy, looking along the main galactic plane. The glow we see is the combined light from many different stars and is visible as a continuous band of light stretching completely around the *celestial sphere*.

Nadir
This is the point on the *celestial sphere* directly opposite the *zenith*.

Nebula
Nebulae are huge interstellar clouds of gas and dust. Observed in other galaxies as well as our own, their collective name is from the Latin *'nebula'* meaning 'mist' or 'vapour', and there are three basic types:

- **Emission nebulae** contain young, hot stars that emit copious amounts of ultra-violet radiation which reacts with the gas in the nebula causing the nebula to shine at visible wavelengths and with a reddish colour characteristic of this type of nebula. In other words, emission nebulae *emit* their own light. A famous example is the Orion Nebula (M42) in the constellation Orion which is visible as a shimmering patch of light a little to the south of the three stars forming the Belt of Orion.

- The stars that exist in and around **reflection nebulae** are not hot enough to actually cause the nebula to give off its own light. Instead, the dust particles within them simply *reflect* the light from these stars. The stars in the Pleiades star cluster (M45) in Taurus are surrounded by reflection nebulosity. Photographs of the Pleiades cluster show the nebulosity as a blue haze, this being the characteristic colour of reflection nebulae.

- **Dark nebulae** are clouds of interstellar matter which contain no stars and whose dust particles simply blot out the light from objects beyond. They neither emit or reflect light and appear as dark patches against the brighter backdrop of stars or nebulosity, taking on the appearance of regions devoid of stars. A good example is the Coal Sack in the constellation Crux, a huge blot of matter obscuring the star clouds of the southern Milky Way.

Neutron Star
This is the remnant of a massive star which has exploded as a *supernova*.

New General Catalogue (NGC)
References such as that for NGC 869 and NGC 884 (in Perseus) and NGC 4755 (in Crux) are derived from their numbers in the New General Catalogue of Nebulae and Clusters of Stars (NGC) first published in 1888 by the Danish astronomer John Louis Emil Dreyer (1852–1926) and which contains details of 7,840 star clusters, nebulae and galaxies.

Occultation
This is the temporary covering up of one celestial object, such as a star, by another, such as the Moon or a planet.

Opposition
Opposition is the point in the orbit of a *superior planet* when it is located directly opposite the Sun in the sky.

Orbit
This is the path of one object around another under the influence of gravity.

Parallax
Parallax describes the change in the apparent direction to a distant object caused by a change in the observer's location. In astronomy, it refers specifically to the very small change in the position of a star when observed from opposite sides of the Earth's orbit. This change, when measured, can be used to infer the distance to the star. The parallax of the nearest star, Proxima Centauri, is 0.768 seconds of arc.

Parsec
A unit of distance, often used by professional astronomers in preference to light years. A star at a distance of one parsec has a *parallax* of one second of arc. It is equal to 3.26 light years. The nearest star, Proxima Centauri, is 1.3 parsecs from the Sun. Distances within our Galaxy are generally expressed in kiloparsecs (1,000 parsecs; abbreviation kpc), whilst distances between galaxies are expressed in megaparsecs (1,000,000 parsecs; abbreviation Mpc).

Penumbra
This is the area of partial shadow around the main cone of shadow cast by the Moon during a solar *eclipse* or Earth during a lunar *eclipse*. The term penumbra is also applied to the lighter and less cool region of a *sunspot*.

Perigee
This is the point in its *orbit* around Earth at which an object is at its closest to Earth.

Perihelion
This is the point in its *orbit* around the Sun at which an object, such as a *planet*, *comet* or *asteroid*, is at its closest to the Sun.

Photosphere
The visible surface of the Sun.

Planet
A planet is a large object that orbits a star and which is made visible by reflecting light from the parent star rather than by producing its own light. There are eight planets in our *Solar System*.

Planetary Nebula
Planetary nebulae consist of material ejected by a star during the latter stages of its evolution. The material thrown off forms a shell of gas surrounding the star whose newly-exposed surface is typically very hot. Planetary nebulae have nothing whatsoever to do with planets. They derive their name from the fact that, when seen through a telescope, some planetary nebulae take on the appearance of luminous discs, resembling a gaseous planet such as Uranus or Neptune. Probably the best known example is the famous Ring Nebula (M57) in Lyra.

Precession
The Earth's axis of rotation is an imaginary line which passes through the North and South Poles of the planet. Extended into space, this line defines the North and South Celestial Poles in the sky. The North *Celestial Pole* currently lies close to Polaris in Ursa Minor (the Little Bear), so the daily rotation of our planet on its axis makes the rest of the stars in the sky appear to travel around Polaris, their paths through the sky being centred on the Pole Star.

However, the position of the north celestial pole is slowly changing, this because of a gradual change in the Earth's axis of rotation. This motion is known as 'precession' and is identical to the behaviour of a spinning top whose axis slowly moves in a cone. Precession is caused by the combined gravitational influences of the Sun and Moon on our planet. Each resulting cycle of the Earth's axis takes around 25,800 years to complete, the net effect of precession being that, over this period, the north (and south) celestial poles trace out large circles around the northern (and southern) sky. This results in slow changes in the apparent locations of the celestial poles. Polaris will be closest to the North Celestial Pole in the year 2102, but it will then begin to move slowly away and eventually relinquish its position as the Pole Star. Vega will take on the role some 11,500 years from now.

Prime Meridian
The celestial prime *meridian* is the meridian on the sky that passes through the *First Point of Aries*. It marks the zero point for measuring *right ascension* and ecliptic *longitude*.

On the surface of the Earth, the prime meridian is the line of constant *longitude* which passes through the centre of the Airy transit telescope at the Royal Observatory at Greenwich in London. It was adopted by international agreement in 1884 as the origin for measuring longitude. Unlike the celestial prime meridian, it has no physical significance.

Prograde Motion
See Direct Motion.

Pulsar
This is a rapidly-spinning neutron star which gives off regular bursts of radiation.

Quadrature
This refers to the geometric configuration of the Sun, Earth and a *superior planet* when the elongation of the planet from the Sun, as seen from Earth, is $90°$.

Quasar
These are small, extremely remote and highly luminous objects which at the cores of active galaxies. They are comprised of a super-massive black hole surrounded by an accretion disk of gas which is falling into the black hole.

Reflection Nebula
See Nebula.

Retrograde Motion
A planet is in retrograde motion when its *right ascension* or ecliptic *longitude* is decreasing with the passing of time. This means that it is moving westwards with respect to the background stars. All *superior planets* undergo a period of retrograde motion around the time of *opposition* and the *inferior planets* undergo a period of retrograde motion around the time of inferior *conjunction*.

Right Ascension
The angular distance, measured eastwards, of a celestial object from the *First Point of Aries*. Right ascension is expressed in hours, minutes and seconds.

Satellite
A satellite is a small object orbiting a larger one.

Seeing
The effects of atmospheric conditions on image quality experienced when carrying out visual observation and astronomical imaging of the night sky.

Shooting Star
The popular name for a *meteor*.

Sidereal Period
The time taken for an object to complete one *orbit* around another, measured with respect to a fixed direction in space.

Solar
Of or appertaining to the Sun.

Solar Eclipse
See Eclipse.

Solar System
The Solar System is the collective description given to the system dominated by the Sun and which embraces all objects that come within its gravitational influence. These include the planets and their satellites and ring systems, minor planets, comets, meteoroids and other interplanetary debris, all of which travel in orbits around our parent star.

Solstice
These are the points on the *ecliptic* at which the Sun is at its maximum angular distance (*declination*) from the *celestial equator*. The term is also used to denote the dates when the Sun passes these points on the ecliptic.

Spectroscope
An instrument used to split the light from a star into its different wavelengths or colours.

Spectroscopic Binary
This is a *binary star* whose components are so close to each other that they cannot be resolved visually and can only be studied through *spectroscopy*.

Spectroscopy
This is the study of the spectra of astronomical objects.

Star
A star is a self-luminous object shining through the release of energy produced by nuclear reactions at its core.

Star Clusters
Although most of the stars that we see in the night sky are scattered randomly throughout the spiral arms of the Galaxy, many are found to be concentrated in relatively compact groups, referred to by astronomers as star clusters. There are two main types of star cluster – open and globular. Open clusters, also known as galactic clusters, are found within the main disc of the Galaxy and have no particularly well-defined shape. Usually made up of young hot stars, over a thousand open clusters are known, their diameters generally being no more than a few tens of light years. They are believed to have formed from vast interstellar gas and dust clouds within our Galaxy and indeed occupy the same regions of the Galaxy as the nebulae. A number of open clusters are visible to the naked eye including Praesepe (M44) in Cancer, the Hyades in Taurus and perhaps the most famous open cluster of all the Pleiades (M45), also in Taurus.
    Globular clusters, as their name suggests, are huge spherical collections of stars. Located in the area of space surrounding the Galaxy, they can have diameters of anything up to several hundred light years and typically contain many thousands of old stars with little or none of the nebulosity seen in open clusters. When seen through a small telescope or binoculars, they take on the appearance of faint, misty balls of greyish light superimposed against the background sky. Although some form of optical aid is usually needed to see globular clusters, there are three famous examples which can be spotted with the naked eye. These are 47 Tucanae in Tucana, Omega Centauri in Centaurus and the Great Hercules Cluster (M13) in Hercules.

Star Colours
When we look up into the night sky the stars appear much the same. Some stars appear brighter than others but, with a few exceptions, they all look white. However, if the stars are looked at more closely, even through a pair of binoculars or a small telescope, some appear to be different colours. A prominent example is the bright orange-red Arcturus in the *constellation* of Boötes, which contrasts sharply with the nearby brilliant white Spica in Virgo. Our own Sun is yellow, as is Capella in Auriga. Procyon, the brightest star in Canis Minor, also has a yellowish tint. To the west of Canis Minor is the constellation of Orion the Hunter, which boasts two of the most conspicuous stars in the whole sky; the bright red Betelgeuse and Rigel, the brilliant blue-white star that marks the Hunter's foot.
    The colour of a star is a good guide to its temperature, the hottest stars being blue and blue-white with surface temperatures of 20,000 K or more. Classed as a yellow dwarf, the Sun is a fairly average star with a temperature of around 6,000 K. Red stars are much cooler still, with surface temperatures of only a few thousand degrees K. Betelgeuse in Orion and Antares in Scorpius are both red giant stars that fall into this category.

Stationary Point
A planet is at a stationary point when its motion with respect to the background stars changes from *direct* (motion) to *retrograde* (motion) or vice versa. All *superior planets* pass through two stationary points at each *apparition*, once before *opposition* and again after opposition.

Sunspot
Sunspots are temporary features on the visible surface of the Sun (the *photosphere*). They appear relatively dark because they are cooler than the surrounding areas of the photosphere.

Superior Planet
A superior planet is a planet that travels around the Sun outside the *orbit* of the Earth.

Supernova
Supernovae are huge stellar explosions involving the destruction of massive stars and resulting in sudden and tremendous brightening of the stars involved.

Synodic Period
The synodic period of a planet is the interval between successive *oppositions* or *conjunctions* of that planet.

Transit
In astronomy, a transit occurs when a relatively small body passes across the disk of a larger body, passing between the body and the observer.

- The passage of Mercury or Venus across the disk of the Sun (as seen from Earth) or of a planetary satellite across the disk of the parent planet.
- The passage of an *exoplanet* across the face of its parent star.
- Another type of transit takes place at the instant when an object crosses the local *meridian*. When the object's *local hour angle* is zero, this is known as upper transit, and marks the maximum *altitude* of the object above the observer's *horizon*. When the object's local hour angle is 12 hours, it is known as lower transit.

Umbra
This is the main cone of shadow cast by the Moon during a solar *eclipse* or Earth during a lunar *eclipse*. The term umbra is also applied to the darkest, coolest region of a *sunspot*.

Variable Stars
A variable star is a star whose brightness varies over a period of time. There are many different types of variable star, although the variations in brightness are basically due either to changes taking place within the star itself or the periodic obscuration, or eclipsing, of one member of a *binary star* by its companion.

Vernal Equinox
The vernal equinox is the point at which the apparent path of the Sun, moving from south to north, crosses the *celestial equator*. In the Earth's northern hemisphere this marks the start of spring, whilst in the southern hemisphere it is the start of autumn.

White Dwarf
A white dwarf is a small, hot star which represents the last stage in the life of stars like the Sun.

Zenith
This is the point on the *celestial sphere* directly above the observer.

Zodiac
The band of *constellations* along which the Sun appears to travel over the course of a year. The Zodiac straddles the *ecliptic* and comprises the 12 constellations Aries, Taurus, Gemini, Cancer, Leo, Virgo, Libra, Scorpius, Sagittarius, Capricornus, Aquarius and Pisces. The ecliptic also passes through part of the constellation of Ophiuchus, as delimited by the boundaries defined by the *International Astronomical Union*, but Ophiuchus is not traditionally considered a constellation of the Zodiac.